Zombies
& Calculus

Zombies
& Calculus

Colin Adams

PRINCETON UNIVERSITY PRESS

PRINCETON AND OXFORD

Jacket image © Jeff Thrower/Shutterstock.
Jacket design by Lorraine Donneker.

Library of Congress Cataloging-in-Publication Data

Adams, Colin Conrad.
Zombies and calculus / Colin Adams.
pages cm

Summary: "How can calculus help you survive the zombie apocalypse? Colin
Adams, humor columnist for the Mathematical Intelligencer and one of
today's most outlandish and entertaining popular math writers, demonstrates
how in this zombie adventure novel. Zombies and Calculus is the account of
Craig Williams, a math professor at a small liberal arts college in New
England, who, in the middle of a calculus class, finds himself suddenly
confronted by a late-arriving student whose hunger is not for knowledge. As
the zombie virus spreads and civilization crumbles, Williams uses calculus to
help his small band of survivors defeat the hordes of the undead. Along the
way, readers learn how to avoid being eaten by taking advantage of the fact
that zombies always point their tangent vector toward their target, and how
to use exponential growth to determine the rate at which the virus is
spreading. Williams also covers topics such as logistic growth, gravitational
acceleration, predator-prey models, pursuit problems, the physics of combat,
and more. With the aid of his story, you too can survive the zombie
onslaught. Featuring easy-to-use appendixes that explain the mathematics
necessary to enjoy the book, Zombies and Calculus is suitable for recent
converts to calculus as well as more advanced readers familiar with
multivariable calculus" — Provided by publisher.

Includes bibliographical references and index.
ISBN 978-0-691-16190-7 (hardback)
1. Zombies—Fiction.   2. Calculus—Fiction.   3. Mathematics—Fiction.
I. Title.
PS3601.D3698Z34 2014
813'.6—dc23
2014012448

British Library Cataloging-in-Publication Data is available

This book has been composed in Minion Pro

Printed on acid-free paper. ∞

Typeset by S R Nova Pvt Ltd, Bangalore, India

Printed in the United States of America

1 3 5 7 9 10 8 6 4 2

Contents

Introduction   1

CHAPTER 1   Hour 6   3
CHAPTER 2   Hour 7   19
CHAPTER 3   Hour $7\frac{1}{2}$ 32
CHAPTER 4   Hour $7\frac{3}{4}$ 48
CHAPTER 5   Hour 8   63
CHAPTER 6   Hour 9   80
CHAPTER 7   Hour 10   95
CHAPTER 8   Hour 18   111
CHAPTER 9   Hour 24   137

Epilogue   152

APPENDIX A
Continuing the Conversations   155

APPENDIX B
A Brief Review of Calculus as   191
Explained to Connor by Ellie

Acknowledgments   223
Bibliography   225
Index   227

Zombies
& Calculus

Introduction

L et me just start this off by saying if you are squeamish, you should not read this book. Of course, there's little likelihood that you are squeamish, considering that you are even reading this at all. The squeamish amongst us, especially those too squeamish to behead a zombie, have long ago been infected and joined the zombie hordes. So if you are reading this, squeamishness is most likely the least of your problems.

This book is my attempt to write down how calculus has helped me to stay alive this long. I am writing this approximately three months after the first incidents of infection occurred. You may be reading this much later than that, and if so, congratulations on still being around. But for me, surviving three months has been a huge victory over immense odds. And honestly, I would not be here today if it were not for calculus. Maybe it will help you to stay alive a little longer, too.

This book is aimed at survivors who have seen some calculus already. I can't teach you all of calculus. That takes a couple of years of work. If you have never seen calculus, you can still read the book, and look at the last appendix, which gives background necessary to understand much of the math in this book, and you will be a better person for it. Perhaps, it will motivate you to study calculus. Of course, these days, trying to stay alive is a full-time job. It's unlikely you will find the time to study calculus as well.

The primary audience for this book is those survivors who learned calculus before the apocalypse. If you have had a year of calculus, that's great. Multivariable calculus is even better, but if you haven't seen that, I'll explain that material as we go along.

Sometimes the conversations I recorded here become more technical. When that is the case, I have moved the more sophisticated parts of the conversation to the first appendix. The places where this occurs are marked by a bloody handprint and a boldface roman numeral, like **(I)**. If you relish the technical stuff, just jump to the appendix and read on as if the text there continues what was being discussed when you encountered the roman numeral, returning to the main text after finishing that section of the appendix. If you don't like technical stuff, just ignore the roman numerals and keep reading straight through the text. Both ways work just fine.

The second appendix is a conversational review of all of calculus, focussing on the aspects that are relevant to this story. If you need some help with the mathematics when it appears in the book, take a look there. It should help.

Life is not what it once was. I have seen things and done things that I would not have believed possible three months ago. But I'm still here, and somehow it makes me feel incrementally better to write this down, to feel like I'm helping others by doing so. This is my story, and in it, I will tell you how you can use calculus to help you stay alive. Good luck to you, and good luck to all of us that remain.

chapter one

HOUR 6

Up until April of this year, I was a math professor at Roberts College, in the tiny town of Westbridge, Massachusetts, on the far western border of the state. Roberts was a small liberal arts college, endowed with excellent students, a beautiful New England campus, and deep pockets. I couldn't imagine a better job than teaching those highly motivated students the mathematics that I love, and also having the time to work on my own research in mathematics. It was a dream job. On top of that, Roberts was set in the Berkshire Mountains, which by anyone else's standards are not mountains at all, but they are still some of the prettiest hills you have ever seen.

Spring break ended at the beginning of April, and we had just returned for the last six weeks of the semester. It was a Friday, and I was reviewing for my calculus class the fact the derivative $f'(x)$ of a function $f(x)$ is the slope of the tangent line to the graph of the function at x. I had drawn this picture (Fig. 1.1) on the board.

Megan, one of those students who always sits in the first row, and who really has a true appreciation for the beauty of mathematics, had asked the perfect set-up question.

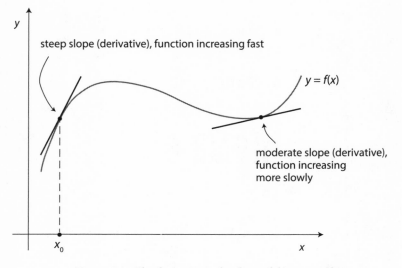

Figure 1.1: The derivative is the slope of the tangent line.

"But Professor Williams, what if the graph isn't smooth? What if it has a kink? How do you know which tangent line to use?"

I smiled. "Great question, Megan. I'm glad you asked." She returned the smile, pleased with the compliment. Megan tended to dress on the conservative side, and that day, she was wearing a tan cashmere sweater over a white blouse, with pearls and a matching tan skirt. Thora, who was sitting just behind Megan, and who had chopped purple hair and more piercings than I could easily count, gave a barely audible groan.

"Suck up," she muttered, just loud enough for Megan and me to hear. I ignored Thora's interjection. In fact, Thora was just as talented as Megan, but she did her best to hide it.

I drew this next picture (Fig. 1.2) on the board.

"At this point, any of these lines could be valid tangent lines. So there's no well-defined slope, and therefore no

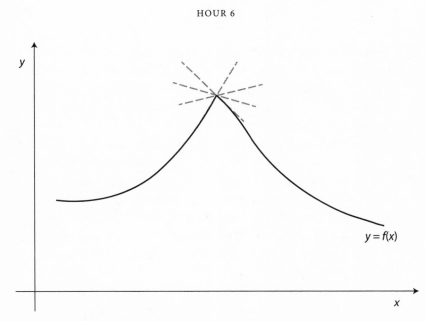

Figure 1.2: A graph with a point where there is no unique tangent line.

well-defined derivative. So the derivative doesn't exist at this point." Megan nodded as I spoke.

It was at this moment that through the window I noticed one of my perennially late students shambling toward class. He was looking particularly disheveled, and I wondered to myself about the appropriate tack to take when he entered the classroom. Charlie wasn't a bad kid, and he wasn't bad at the math either. He was just a serious flake who was too disorganized to attend class regularly or turn homework in when it was due. If I had to guess, I suspected the reason he was late this Friday was a drunken binge the night before. Thursday night parties were becoming more and more of a problem.

Although I anticipated the door opening, I continued on. "At all the other points on the graph other than the kink

5

point, there is a perfectly good tangent line, and therefore, everywhere else the derivative is defined." Charlie still didn't enter. This distracted me to the extent that I finally crossed to the door and opened it. Charlie was standing in the lobby facing away from me.

"Are you going to join us, Charlie?" I asked.

He turned to look at me, his expression suggesting he didn't register who I was. But then his lips pulled back into what might have been a grimace and what might have been a bizarre smile. There was definitely something off-kilter about him. His eyes looked filmy, and his posture seemed all wrong. Spittle hung off his chin. That must have been some party.

He shuffled forward and I backed into the room, not sure what to make of him. He stopped when he saw row after row of students staring at him. I tried to regain my composure.

"Nice of you to join us, Charlie. Perhaps you would like to take a seat." I motioned to the front row.

Charlie lurched forward and instead of taking the seat I had intended, he fell on top of Megan. Initially I and all the other students assumed Charlie had tripped, and there was a broad giggle from the class. But almost immediately, Megan started shrieking, and suddenly blood was spurting out. Charlie made guttural snarling noises as Megan flailed about. It took me a moment to register what was happening, but then I leaped forward.

"Get off her!" I screamed. Grabbing at Charlie's arm, I tried to pull him back, but his teeth were embedded deep in Megan's neck. I pulled harder, putting all my weight into it, and the three of us toppled back onto the floor, with me on the bottom and Megan on the top. Charlie never released his grip as he climbed back on top of her and continued to chew on her neck.

Tom and Manuel, two of the biggest students in the class, both members of the front line of the football team, leaped up from their seats at the front of the room. They tended to sit near the door to make their escape as soon as possible. Tom grabbed Charlie around the waist and Manuel held onto Megan as Tom tried to pull Charlie away. There was a ripping sound and much of Megan's neck came off in Charlie's mouth. By now there was lots of screaming.

Charlie then turned around as Tom was struggling with him and bit Tom's nose off. At this point, people started bolting for the door, knocking each other over in an attempt to get out. Tom fell over backward screaming, and Charlie fell upon him.

"Get out, everyone get out," I cried over the tumult. Shrieking students scrambled out the door. Thora sat frozen in her seat. I grabbed her by the arm and pulled her up. "Run!" I screamed, as I pushed her through the door ahead of me.

Charlie continued to feast on Tom. I slammed the door behind me, and ran across the lobby to the department's administrative assistant's office. Having heard the commotion, Marsha was standing in her doorway. She was dressed in her usual color-coordinated manner. Today it was red high heels, red dangly earrings, red lipstick, and a short red skirt with matching red stockings.

"What's going on, Craig? Give back an exam today?"

I pulled her into her office and shut the door. "Call security! Call the Westbridge police! Call a SWAT team from somewhere! A student's gone completely insane. He's killed two students."

Marsha laughed. "It's a little late for April Fool's."

"No April Fool's," I said, handing her the phone. "Call! Call now!" She could tell by the tone of my voice that I wasn't kidding and she began to dial. I peeked out the door, but the lobby remained empty. All the students from my class had gotten as far away as possible. I realized I had to clear the building as quickly as I could. Cautiously, I stepped out into the lobby and pulled the fire alarm on the wall. As the alarm sounded, flashing blue lights strobed on and off in the lobby. Almost immediately students and faculty poured out of the classrooms into the lobby.

"Get away from the building," I yelled. "This is a real emergency. Get far away from the building." Within five minutes, the lobby was again empty. I then cautiously crossed to the door of the classroom to check that it was still latched shut. I could hear slurping sounds emanating from inside. Slipping out the front door of the building and keeping low, I snuck around the outside to the window I had looked out when I had first spotted Charlie. I peered in, and saw Charlie still chomping on Tom, blood oozing down his jaw.

I reentered the lobby as Marsha came out of her office.

"Security and cops are on the way," she said.

"Thank God," I said.

"What's going on in there?" she asked.

"You don't want to know," I replied.

"Actually, I do want to know," she said.

Before I could explain, we heard someone coming down the hall that connected the back door of the building to the lobby. It was Hoyle, the Westbridge police chief. Stuffed into his uniform like a sausage in its casing, he had a perpetually angry expression on his face. I had never been a particular favorite of his since the time I stood up at a town meeting and argued that we should cut the town's police budget. I didn't see the point of having three on-duty police officers for a town that averaged one crime every six months.

"What the hell is going on?" demanded Hoyle over the fire alarm. He ignored me and turned to Marsha. "Did you make that call?"

Marsha nodded and pointed at me.

"There's a crazy student in 106," I said as calmly as I could muster.

"So call Psych Services. Why're you calling me?"

"Umm, because he's killed two students."

"You're the one who sounds crazy right now," said Hoyle. "Did you pull the fire alarm?"

"Yes," I replied. "I had to get everyone out of the building."

He walked over to the alarm box, inserted a key and silenced the alarm. Then he walked over to the door to the classroom.

"Really!" I said. "He bit them, and then started eating them."

"A little late for April Fool's isn't it?" he replied, as he pulled out his gun, and reached for the door handle.

"I wouldn't do that," I said, as he pushed open the door.

From my vantage point I could see blood coating the linoleum. I could also see Charlie kneeling over Tom's prone form. But the sight that was the most disturbing was Megan standing in front of her desk. Her head was tilted at a funny angle due to the chunk of her neck that was missing, but she was definitely up and about. She saw Hoyle, gave out a guttural moan, and immediately moved for him. Lucky for him, as she stepped forward, her foot slipped in the blood on the floor, and she went down hard, her head bouncing on the tile.

Charlie looked up at this point and realized a better meal had presented itself. As he rose from the floor, I could see Hoyle fiddling with the safety on his gun. Charlie reached

9

him just as the gun went off. Both Charlie and Hoyle went down.

"Oh, shit," I mumbled. I could see Megan struggling to get up, as Charlie sank his teeth into Hoyle's leg.

"Holy crap," cried Hoyle. "Help me!" He reached imploringly in my direction.

Hoyle's gun lay on the floor just out of his reach.

"Don't," said Marsha, but I couldn't just watch the rest of this scene unfold. I ran across the lobby and grabbed up the gun. Pointing it at Charlie's chest, I pulled the trigger. There was a loud explosion and the force of the bullet knocked him flat on his back. Hoyle started to crawl away. As I reached down to help him, Charlie inexplicably lifted his head off the tile. He seemed unfazed by the gaping wound in his chest as he reached out and grabbed at Hoyle again.

"What the hell," I uttered incredulously.

"Shoot him again," screamed Hoyle. I lined the gun up on the side of Charlie's face and pulled the trigger. Brains flew out the other side of his head, and he collapsed. Hoyle crawled out from under him and pulled himself through the doorway. I just stood there with the gun held limply in my hand. In the meantime, Megan had managed to regain her footing. She slowly stepped forward, seemingly aware that a fast movement would put her back on the floor. As she was reaching for us, I grabbed hold of the door handle and slammed the door shut. I was shaking uncontrollably. I had never shot a gun before, let alone at a student, crazed or not.

"Jesus Christ, that was some messed up student," said Hoyle wincing. "Help me up."

Hoyle wrapped an arm over my shoulder and I hoisted him to his feet. He motioned for his gun, which he reholstered. Then, as I supported him, he used his shoulder radio to call the police station.

"Chuck, pick up. We need serious backup ... Chuck. Come on. Pick the hell up. Where the goddamn hell are you?" He clicked off his radio in frustration.

"There's always supposed to be someone there," he said to me. With my support, he hobbled across the lobby.

Marsha motioned to the glass door of the lobby.

"Thank God. Here's campus security," she said. I saw Hollister, one of the college security guards, heading toward the building. Marsha clattered on her high heels to the door and pushed it open.

"Hurry!" she yelled as she held the door for him. At that moment, I noticed that Hollister's gait looked peculiar. He was not bending at the knee as he shuffled forward.

"Marsha, let the door go," I screamed as I pointed at Hollister. As he was reaching a crabbed hand for her, she suddenly grasped the reality of the situation and ducked back into the building, slamming the door. Hollister's face smashed up against the glass as he flailed about, desperate to sink his teeth into her.

Marsha sank back, horrified.

"What the hell is going on?" I asked to no one in particular, staring as Hollister continued to ram his face into the glass door.

"Oh, my God," whispered Marsha in horror. We watched as he tried to bite through the smooth surface, his saliva smearing the glass.

"We have to get out of here," said Marsha.

"Not sure that's such a good idea," I said.

"What do you mean?" demanded Hoyle. "Of course we have to get out of here."

"Look," I said. "In the last fifteen minutes, three people have either tried or succeeded in killing and eating others."

"So, that seems to be an argument for getting as far away as possible," said Marsha.

"Actually, it's not," I replied. "The point is that if two of the three people who have tried to enter the building are crazy, it is very unlikely that they are the only two who are crazy. But if a whole lot of people are crazy, then it's not so surprising that two who tried to come in are crazy. It's probability."

"So you're trying to say that there are lots of others who are nuts?" asked Hoyle. "And if we go outside, our chances of survival plummet?"

"That's exactly what I'm trying to tell you," I said. "Come on. We can check it out from the windows upstairs."

With Marsha on one side and me on the other, we managed to help Hoyle up the stairs to the second floor. As I pushed open the fire door to the stairwell, we collided with Angus. He was a student who had taken Calculus I with me two years previous and had somehow managed to pass the class, just barely. For some unknown reason, that had convinced him to be a math major.

"What's going on?" he asked. He used to ask that a lot in class, too. "I heard the fire alarm and a lot of screaming."

"We don't know exactly, but people are acting crazy," I replied.

"Very crazy," added Marsha.

From down the hall, I saw a head poke out an office door. It was Jessie, a professor in the Biology Department with a strong mathematical bent. She had been my wife's best friend, and my wife had helped her through a difficult divorce. When my wife was diagnosed with breast cancer, Jessie was there for us through the entire two-year ordeal, which culminated in my wife's death five years ago. After that, Jessie did her best to console me and help with my kids, and that led very quickly to an ill-conceived attempt at a relationship. But it had been much too soon for me. The relationship disintegrated quickly and after that we had avoided each other. But under the circumstances, these personal issues seemed suddenly peripheral.

"Craig, get down here," she called. With Angus and me helping Hoyle, we made it down the hall. Jessie locked the door once we were inside. Sitting in Jessie's office chair was Oscar Gunderson.

"Hello, Williams," he said. Those were the first words he had spoken to me in eight years.

Gunderson was tall and thin, with a shaved head and thick-rimmed, stylish, chunky, black glasses that looked out of place on his long face. He sported the carefully trimmed stubble that always looked to be much more trouble than the careless attitude it was supposed to project. His clothes always conformed to whatever *Men's Style* announced as the latest fashion.

Gunderson had arrived at Roberts College three years before me. As the hot young applied mathematician with a National Science Foundation grant to support his research and invitations to speak all over the world, he had cruised through the tenure process. Then, three years later, when I came up for tenure, I barely scraped by, as I was to learn afterward. I had plenty of papers, both single author and with my students. But Gunderson had been on my tenure review committee. He argued that my chosen field of low dimensional topology was dead, and that it would be a big mistake to tenure someone in an area that would soon cease to exist. He also argued that the fact I was publishing papers with students meant it couldn't be good work. Lucky for me, there had been other senior faculty in the department who believed in me, and I had received tenure in spite of his best efforts. Two years later, Perelman proved the Poincaré Conjecture using some of my work on Ricci flow. So my supporters were vindicated and I had plenty to work on for the next half century. I ignored Gunderson as we lowered Hoyle into a chair.

"Jessie, do you know what the hell is going on?" I asked. She motioned to her window, which gave a good view of the science quad.

There was mayhem below. Hollister had obviously given up on getting into the Science Center, and we watched as he managed to grab hold of a student's hair as she ran past. He jerked her off her feet and fell upon her as she screamed desperately. We saw several other people feasting on bodies. Various faculty and students wandered around the quad with that identifiable stilted gait. When an uninfected person ran past, they would turn after them in pursuit. Most of the runners would escape to safety, but once in a while someone would trip, or dodge one infected person only to stumble into another.

"I checked on the internet," said Jessie. "It's a virus of some kind. Spreading fast. Started in Boston, maybe at one of the university labs, but no one knows for sure. It shuts down all your higher brain functions, everything we have evolved over the last fifty million years. All that's left is the basics."

"What basics?" asked Marsha.

"Breathing, and the need to eat. Those functions have been around all the way back to the very beginning."

"They're zombies," said Angus.

"They are not zombies," said Jessie. "They are sick human beings."

"What's the difference between them being zombies and them being sick human beings?" asked Marsha.

14

"Zombies are in the movies," said Jessie.

"This is like a movie," said Marsha, "a horror movie."

"It's the Z virus," said Angus."Probably coming out of research funded by the military. They're always trying to create a superweapon."

"How do you know that?" asked Gunderson.

"Everybody knows it," said Angus. "It's the basis for half the sci-fi movies out there."

Jessie smiled. "You know, it might not be a conspiracy. Might just be some scientists trying to find a cure for an existing virus, and they messed up."

"Does it matter which it is?" asked Gunderson.

"This is all very messed up," I said. "I have to call my kids." I pulled my cell phone from my pocket and dialed my seventeen-year-old daughter Ellie's cell. I waited three rings anxiously and then, much to my relief, she answered.

"Honey, are you okay?"

"Yeah, Dad, but something crazy is going on. They sent us all home from school an hour ago. Said there was some virus, and everyone should go home. I tried to call you but you must have had your phone off because you were teaching. People were acting crazy when the bus went by campus. What's going on?"

"It's nuts, honey. Is your brother there?"

"Yeah. He got sent home, too."

That was a relief. Ellie was very sensible and trustworthy in an emergency. Her twelve-year-old brother Connor was anything but.

"Okay, look. I want the two of you to lock every door of the house. Don't let anyone in. Then go upstairs and lock yourselves in the master bedroom. Bring some snacks and drinks with you. And the dog. And don't leave. Got it?"

"Dad, you're scaring me."

15

"You should be scared, honey. It's very dangerous. I'll get there as soon as I can."

"Hurry, Dad."

I hung up. Angus was staring out the window.

"Look at the way they move," he said. "Very stiff. And they all seem to move at the same speed."

"True," I said. I watched as Travis McCutcheon, star of the track team, lurched across the quad at the same pace as Reverend Mitchell, the seventy-year-old college chaplain.

"Looks like they move at about a yard a second," said Angus.

"A constant derivative," said Gunderson.

"What do you mean?" asked Angus.

"Haven't you taken calculus?" asked Gunderson.

"Sure," replied Angus.

"Who did you have?" asked Gunderson. Angus turned and pointed at me.

"Figures," said Gunderson.

"Angus, you do know this," I said. "The derivative of a function is just the rate of change of the function, which we measure by how steep the tangent line is. So if a function is changing fast, it has a high derivative and if it is changing slow, it has a low derivative."

"Yeah," said Angus. "That sounds right."

"If the function happens to be the distance you have travelled in a certain time," I continued, "then the derivative is just the rate of change of the distance travelled, which is your speed."

"I thought speed was distance divided by time," said Marsha.

"That's right," I replied, "if you're moving at a constant speed. If you're going 50 miles an hour for an hour, you go 50 miles in that hour. But you could also go 50 miles

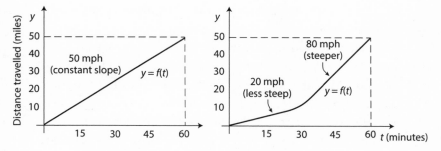

Figure 1.3: Two ways to travel 50 miles in an hour.

in an hour by going 20 miles per hour for half of the time and then speeding up to 80 miles per hour for the rest of the time." I drew Figure 1.3 on the white board hanging on the wall.

"Oh yeah," said Angus. "Your speed usually varies. So when you're going 20 miles per hour, your derivative is 20 mph and when you are going 80 miles per hour, your derivative has increased to 80 mph. Like when you start running. You start at a slower speed but then you build up to a faster speed. So your derivative changes. Like Doctor Ortiz there."

Raphael Ortiz, a faculty member in chemistry, had been hiding behind a bush, his orange polo shirt barely visible. But he had seen an opening, so he made a break for it. It took him a second to get up to speed, but I could see he was in good shape as he sprinted toward the parking lot. He attracted several zombies, including McCutcheon and the chaplain, but he had a good head of steam and had no trouble outdistancing them. As Angus had pointed out, he started slower but got going faster until he reached his top speed, which he maintained for most of the distance. He had strong incentive.

17

Ortiz easily made it to his car but then took a second fumbling with his keys as the zombies converged toward him. He managed to get the door open before they arrived, and slammed it shut just as the first reached him. Within seconds, McCutcheon, the chaplain, and two other zombies surrounded the car, banging on it as they slobbered over the windows. Ortiz got the car started and jerked the car backward out of its parking space, causing the chaplain to fall in front of the car. Ortiz then slapped it into drive and floored it. The car immediately bumped over the chaplain and sideswiped another car. At this point Ortiz realized that an abandoned car blocked the gate to the parking lot.

He swung his car around the other side of the lot and gunned it over the curb onto the quad itself. As he continued to accelerate, his tires chewed two big tracks into the college's carefully manicured lawn. I couldn't help but think of my neighbor Winsted, who was the head of Buildings and Grounds. He would not be happy about this.

As the car cut a swath across the quad, all of the rest of the zombies in the quad were distracted from their assorted atrocities and headed toward it. Ortiz didn't seem too concerned for them as they lurched in front of the car. He mowed down three in a row. But when the next one to step in front of his car was Karen Holm, chair of the Chemistry Department and, at least according to the rumor mill, his lover, he swerved to miss her and plowed into the big oak in the center of the quad. The front end of the car crumpled as the air bag deployed.

All the zombies now converged on the car, but as far as I could tell, the glass had not shattered and the passenger compartment was intact. Steam rose from the cracked radiator. Ortiz wasn't moving.

chapter two

HOUR 7

"That poor man," said Marsha.

"We have to help him," said Angus. "I had him for Physical Chemistry. He gave me a C."

I didn't know if that was good or bad as far as Angus was concerned.

"There's not a lot we can do," I said. "He should be okay for now. I don't think they can break the glass and they certainly can't open car doors."

"Okay," said Jessie. "We can worry about Ortiz later. Right now, if we're going to survive, we need a plan."

"Call the National Guard," said Hoyle, grimacing from the pain in his leg. "Get some ordinance in here."

"I don't think the National Guard will be arriving in Westbridge in the near future," said Jessie. "The entire state is a disaster, and it sounds like it's spreading fast."

"How fast?" I asked.

"According to a model I found on the Web, it's exponential growth," she replied.

"What does that mean?" asked Marsha.

"Exponential growth means that the number of people infected is growing very fast," I answered.

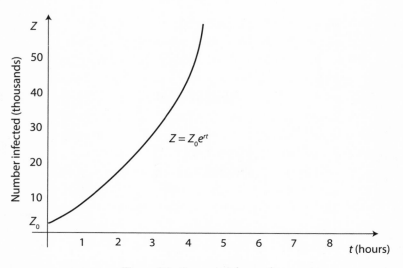

Figure 2.1: Exponential growth.

"How fast?" asked Angus.

"Exponentially," replied Gunderson in a sarcastic tone.

"It means," I said to Angus, "that the number of zombies is growing like the exponential function. That is to say, if Z is the population of zombies at any given time t, then we have a function like this." I wrote this equation on the board.

$$Z = Z_0 e^{rt}$$

Then I drew Figure 2.1.

"Okay," said Marsha. "I'm good with the Z and the equal sign. But I don't get the Z_0, the e, the r, and the t."

"Is that all?" said Gunderson.

"Ignore him," I said to Marsha. "Z_0 is the size of the initial population of zombies. We can probably assume this started with one individual who was somehow infected. So we can take $Z_0 = 1$."

"Okay," said Marsha hesitantly.

"The e is the number e, approximately 2.71828."

"Approximately? Approximately would be 2.7 or 3."

"Well, we actually know this number out to a lot of decimal places. It's a number that turns up a lot in mathematics."

"I know about e. I've heard it come up in discussions between math faculty before," said Marsha. "You assume I'm not listening but in fact I am."

"I wouldn't say we assume you're not listening, " I said.

"This is irrelevant," said Gunderson. "Who cares?"

"Okay," I said. "Now t is the time that has passed since the infection started, let's say in hours. Jessie, did it say anything about when the infection might have begun?"

"Yes. By now it's been about six hours since the first case was reported. Supposedly some janitor at Harvard tried to eat a freshman."

"I'm surprised we heard about that," said Gunderson. "I would have thought the Harvard administration would have tried to keep that hush-hush. A story like that can't help the alumni fundraising campaign."

"Well, actually, I think they did try to keep it hush-hush," said Jessie. "But when the student then turned around and bit a finger off a paramedic, the story took on a life of its own."

"Hmm, janitor at Harvard. Gives credence to the theory it started in a lab," I added.

"So what's r?" asked Angus.

"That's the growth rate," I replied. "It's determined by the number of new infections caused by a single individual in a single hour."

"So what's that?" asked Marsha.

"I don't know. But different growth rates generate very different functions." I drew Figure 2.2.

21

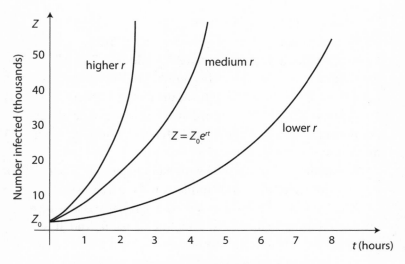

Figure 2.2: Growth rates yield different functions.

"Let's suppose each zombie infects five other people per hour. Then, the population of zombies quintuples every hour. To get that, we let $r = \ln 5$."

"Really? Why ln 5? That seems like a funny choice," said Angus.

"Because then the function becomes $Z = Z_0 e^{\ln 5t}$. Since e^x and $\ln x$ are inverse functions, they cancel each other out and we get

$$Z = Z_0 e^{\ln 5t} = Z_0 5^t.$$

So each time we increase t by an hour, the population of zombies quintuples. The current zombie population would then be $Z = Z_0 5^t = 1(5^6)$." I took the calculator off Jessie's desk, and punched in the numbers.

"That's about 15,000 infected individuals."

"Wow!" said Marsha.

"And that's a conservative estimate. If we just assume six people per hour, then we get $e^{\ln 6(6)} = 6^6$, which is," I said while punching the calculator, "a grand total of about 45,000. In two more hours, it will be over one and a half million."

"Now, that's a lot of zombies," said Marsha.

"Hey, I remember seeing a TV show where they doubled the number of pennies every ten seconds," said Angus. "Within six minutes there were more pennies than there are people on earth."

"What kind of TV do you watch?" asked Marsha.

"It was on public television," said Angus with a shrug. He paused for a moment and then said, "But I guess I don't understand. With this exponential model, it'll quickly grow to be bigger than the number of people on earth."

"You're right, Angus. This is only an initial model, when the number of noninfected is much greater than the number of infected. Once the number of infected becomes closer to the number uninfected, the model no longer holds."

"The great part of being a math teacher must be that you get to make up math," said Marsha.

Gunderson laughed. "Yeah, Williams. Good job making up math."

"I don't make it up, Marsha," I said. "There's a reason it's exponential growth."

"What's that?" asked Angus.

"It has to do with a differential equation."

"What's a differential equation?" asked Marsha.

"It's just an equation that has derivatives in it."

"Then it should be called a derivative equation," said Marsha. "Wouldn't that be a better name? Less confusing?"

Gunderson snorted.

"Maybe," I replied. "But derivative and differential are essentially synonymous."

" 'Derivative' is the noun," said Gunderson, "and 'differential' is the adjective. Basic grammar."

"But isn't the word 'differential' also used for dx or dy?" asked Angus.

"Oh, no," said Gunderson, "a single word with more than one meaning. The English language will crumble."

Angus clenched his fists, his body language betraying his reaction. Marsha put a calming hand on his shoulder.

"Just ignore him," I said. "In this case, the differential equation is $\frac{dZ}{dt} = rZ$."

"Why would that hold?" asked Angus.

"It only holds initially, but assume, for instance, that there is a population of mice and they have an unlimited food supply, and no predators. Then the size of the population is only limited by how often they can procreate. Mice are known to have a litter of ten every month. And then the litters have a litter, et cetera. Two mice can generate a population of a million in six generations."

(I. Growth continued on p. 155)

"But then what about the differential equation?" reiterated Angus. "What does this r stand for? Is it the same as the r in the original equation you showed us?"

"Yup, same as before. That's the growth rate for the population."

"But how does any of that apply to the zombies?"

"Well, think of the population of zombies. For all intents and purposes, at least initially, they have an unlimited food supply. There are over three hundred million humans in the U.S. alone."

"Yes, but unlike mice, the zombies can't have babies," said Marsha. "At least I hope not."

Angus laughed. "Imagine two zombies having sex. Yuck!"

Jessie shrugged. "Hate to mention it, but sex has been around a very long time. It's a lower brain function. So who knows?"

"Don't even go there," I said. "But as far as we know, the population only grows by converting the uninfected humans. Because there's currently an essentially limitless supply of uninfected humans, the factor that determines how much contact there is and therefore the rate of growth of the population is the size of the population itself. So the rate of change of the size of the population of zombies is, at least initially, proportional to the current number of zombies. The more zombies there are, the faster the number will grow."

"But what determines r?"

"It would depend on how often a zombie comes into contact with a human, and when the two come into contact, how often the zombie infects the human."

"Sometimes, the human could kill the zombie," added Marsha. "You killed that one who was trying to eat the chief. We don't always lose the battle."

"True, " I said. "So r will also be affected by that."

"But what does that differential equation have to do with the other equation you wrote?" asked Angus. "The $Z = Z_0 e^{rt}$?"

"$Z = Z_0 e^{rt}$ is a function that satisfies the differential equation. So it's the function that models how the number of zombies grows, at least initially."

"What do you mean, it satisfies the equation?" asked Marsha.

"If you replace Z in the differential equation with $Z_0 e^{rt}$, the equation will hold. Look, let's take the derivative of $Z = Z_0 e^{rt}$. Angus, do you remember how to do that?"

"This should be fun," said Gunderson.

"I remember the derivative of e^x is e^x," said Angus. "It's the only function that is its own derivative. You used to say it's the only function that gives birth to itself, making it its own mother. That stuck with me. The image of someone giving birth to themselves. Hard to forget that."

"That's right, Angus," I said, nodding. Gunderson rolled his eyes.

"So then, the derivative of $Z_0 e^{rt}$, umm, how do I take the derivative of Z_0?"

"It's just a constant. Do you remember what happens to constants when you take the derivative?"

"Oh, yeah. They pass right through the derivative, so you get this." He wrote on the board:

$$\frac{d}{dt}(Z_0 e^{rt}) = Z_0 \frac{d}{dt}(e^{rt}).$$

"Then I think I use the Chain Rule, right?" he continued. "Derivative of the outside function at the inside function times the derivative of the inside function. The outside function is e^x, really, so I jut get $Z_0 e^{rt}$ times the derivative of the inside function. The inside function is rt, so its derivative is r. So the derivative is $Z_0 e^{rt}$ times r."

"Very good, Angus," I said. "So you showed that when $Z = Z_0 e^{rt}$, then $\frac{dZ}{dt} = Z_0 e^{rt} r$. And what is that, in terms of Z?"

"Well, since Z is $Z_0 e^{rt}$, that's Zr. So we found that when $Z = Z_0 e^{rt}$, then $\frac{dZ}{dt} = rZ$, which is what we wanted to show."

"Exactly!" I said, smiling triumphantly. "So that's why exponential growth models the number of zombies, at least initially."

Gunderson waved dismissively. "I can teach a parrot to repeat what I say, too."

"I'm not a parrot," said Angus indignantly. "I understand it."

At this point, Hoyle moaned. "My leg really hurts," he said.

"Let's take a look," said Jessie. She tore the already ripped pant leg out of the way, exposing the open wound. A decent chunk of flesh was missing, and blood was running down the leg. Jessie grabbed a few tissues from a Kleenex box on her desk and swabbed away the blood. The surrounding tissue was a bright red.

"What happened?" asked Jessie. She took a scarf from around her neck and tied it tightly around Hoyle's calf.

"One of those damned zombies bit me," replied Hoyle.

"Oh," said Jessie. She caught my eye and then leaned over to me and whispered, "We need to talk . . . in private."

"Really?" I said. "Is that necessary?" I was thinking about the potential danger lurking in the hall.

"It is," she replied.

"Okay," I said, motioning to the door.

"I'm coming, too," said Gunderson.

"Me, too," said Marsha. Angus continued to stare out the window.

"Okay," I said. I opened the door to the office slowly and peeked out. The hall was empty. Jessie ushered the four of us out of her office and shut the door behind her.

"Craig, he's been bitten."

"So?"

"So he'll turn. That's how the virus is transmitted. Through a bite or a scratch. From their bodily fluids to ours." I remembered the image of Megan standing with her head at that crazy angle.

27

"How long will it take?" asked Marsha.

"We only know one example," I said.

"Who's that?" asked Gunderson.

"Megan Verbedo, my student in calculus."

"I know her," said Gunderson. "She's got a lot of potential."

"Not anymore she doesn't," I said. "She was bitten by Charlie, who was infected, and then she rose from the dead."

"They don't actually rise from the dead," said Jessie.

"Sure they do," said Marsha. "Everybody knows that."

"Just in the stories, they do," said Jessie. "Fact is they are living, functioning human beings, just missing all higher brain function."

"But I shot Charlie in the chest," I said, "and he still kept coming."

"Yes," replied Jessie. "But that's because he had no cognizance of having been shot in the chest. Didn't have the brain function to understand it. Often, when someone is shot, they are incapacitated by the knowledge they have been shot and the pain associated with it. No knowledge and no pain, well then they can do superhuman things. But in any case, you must have missed his heart. That would have stopped him dead. There is no animal with a heart that can keep coming once you destroy it."

"I'm pretty sure you have to destroy their brain," said Marsha.

"Forget the movies, Marsha," said Gunderson. "This is the real world."

"The heart will do the trick," said Jessie. "Or the head. And even in missing the heart, the zombie still would have bled to death, but until he did, he would continue to try to eat you and do exactly what he was doing before he was shot."

"What about Megan?" I asked. "She was missing a big chunk out of her neck."

"I'm guessing that she was not missing her carotid arteries, because otherwise she would have bled out fast, and couldn't have been walking around. There's nothing supernatural here. These are flesh-and-blood creatures, just like us, only they will keep on coming until you put them down permanently."

"But why wasn't more blood coming out of her neck?" I asked.

"Maybe the virus generates a natural coagulant. Would make some sense, if it evolved. Then the host is less likely to perish fast, meaning there's more time to pass the virus on to others."

"If these infected people are still people, instead of walking dead people, maybe we shouldn't kill them," said Marsha. "Maybe someone will come up with a cure."

"I doubt it," said Jessie. "According to what I read online, the virus doesn't just shut down brain function. It actually liquefies the parts of the brain that contain that function. There's no reversing it once it starts."

"Yuck," said Marsha.

"So how long from first bite to full infection for Megan?" asked Jessie.

"No more than fifteen minutes," I replied.

"And how long has it been since Hoyle was bitten?"

"About the same." The four of us traded glances.

"Should we warn Angus?" suggested Marsha. "He's in there with him."

"Hoyle's a big fella," I said, "A lot bigger than Megan. If the infection time is proportional to weight, then Hoyle weighs at least twice as much as Megan, so it should take twice as long for him to become a zombie."

Jessie shrugged. "It could just as easily be a square root relation."

"You're talking gobbledygook again," interjected Marsha.

"Craig was saying that maybe the infection time is proportional to weight. So if someone weighs w, the infection time t is k times w. So we write $t = kw$. Double the weight on the right side of the equation, and that doubles the time until infection."

"That makes sense," said Marsha.

"But I was suggesting that the formula could just as easily be something like $t = k\sqrt{w}$. Then doubling w only multiplies the infection time by $\sqrt{2}$. So if Megan took fifteen minutes to turn, then Hoyle might take twenty-one minutes."

"Maybe it will take even longer," said Marsha.

"Maybe," said Jessie, "but we have to deal with this now." Down the hall I suddenly spotted a student walking toward the stairwell door. She had on headphones and seemed oblivious to the crisis unfolding around her.

"Hey," I yelled as I waved at her. She glanced at me, smiled, and waved back just as she was pushing the stairwell door open. I saw a hand reach out and grab her by the shirt as a startled look crossed her face, and then she went down. Her legs were still sticking out of the stairwell, jerking as she started screaming.

"Quick," I yelled. "Back into the office."

The door to the stairwell pushed further open and several zombies fell through, tumbling over the student's body. They saw us and immediately scrambled to their feet, staggering in our direction. Jessie unlocked the door to her office and we slipped inside as I saw a half dozen hobbling down the hall. As we slammed the door, we could hear them moaning and scrabbling on the outside.

"What the hell?" said Hoyle. His face was red and sweat was dripping off the end of his nose.

"Bit of a problem," I replied.

"Bit?" said Gunderson. "We're now trapped in here with a bunch of zombies at the door and another one brewing on the inside."

"What do you mean one brewing on the inside?" asked Angus. Hoyle let out a groan.

chapter three

HOUR 7½

Outside the office door, we could hear the zombies moaning and banging into one another, frustrated by the proximity of their potential meal.

"What do you have for weapons in here?" I asked.

Jessie looked around her office. "I don't really keep weapons in my office," she replied.

Marsha picked up a stapler. "I'm pretty good with this," she said. Gunderson snorted.

"We can use Hoyle's gun," said Jessie quietly to me.

"First we have to get it from him," I replied.

"Why don't you ask him for it?"

"It might go better if you asked," I answered.

She frowned at me and then turned to him. "Chief Hoyle, can I have your gun? We have a bit of a zombie infestation just outside the door."

"Nobody touches my gun," said Hoyle.

"I already touched it," I said. "In fact, I used it to save your ass."

Hoyle put his hand on the butt of the gun.

"Nobody, I repeat, nobody touches this gun."

"You're touching it now," said Angus.

Hoyle glared at him.

"What else could we use?" asked Jessie.

"Here are some scissors," said Marsha.

"We need something heavy," I said.

"We could stab them in the eye with a pen," said Marsha.

"I think that's a long shot," said Jessie.

"What about my stockings?" said Marsha.

Gunderson laughed. "You're kidding, right?" he said.

"No," said Marsha. "We could put something heavy in them, and swing it around."

"You know, I think that might work," I said.

"I'll take them off. Don't look."

I turned my back. Angus did likewise. Hoyle was in too much pain to notice or care. Gunderson crossed his arms and continued to stare at her.

"Oscar," said Jessie sharply. "Turn around."

Gunderson rolled his eyes and then did so.

Marsha removed her stockings and handed them to me. Taking the scissors on Jessie's desk, I cut them to separate the two legs. Tying a knot in the toe of each to minimize the likelihood of ripping, I then dropped a heavy coffee cup in one and a crystal paperweight in the other. It was the paperweight that faculty received from the college upon being granted tenure, with the college logo embossed on the top.

"Just get a good swing up. Should do some damage," I said, as I handed one to Jessie and one to Angus.

"Are you kidding?" said Gunderson. "You're telling them to defend themselves from a raving lunatic that wants to chew on their face with a paperweight in a stocking?"

"Oscar, you're not helping," said Jessie. "We're trying to come up with ideas here."

"But that's going to get us killed," said Gunderson.

"Gunderson," I said. "You swing that paperweight at one revolution per second, and with a radius of 1 meter, then the circumference of that circular path is $2\pi r$. That's 2π meters. So the paperweight travels 2π meters in a second. That means its speed is 2π meters per second."

"So?"

"So I would estimate that the paperweight weighs a kilogram, which is about 2.2 pounds. Then we can use the physics equation $F = ma$ to figure out the force F."

"What's a?" asked Angus.

"That's the acceleration."

"I thought acceleration was the derivative of velocity, which is the derivative of the position function."

He wrote on the board:

$$\text{position at time } t = f(t),$$

$$\text{velocity } v(t) = \frac{df}{dt},$$

$$\text{acceleration } a(t) = \frac{dv}{dt} = \frac{d^2 f}{dt^2}.$$

"Acceleration is the second derivative of the original function that gives your position," he said.

"That's all true, Angus, but in this case, the acceleration means the change in velocity of the paperweight when it comes into contact with the head."

"But how would we know that?"

"We can figure out the average acceleration when the paperweight comes into contact with the head, which is the average change, by just taking the initial velocity minus the final velocity and dividing by the time interval Δt that the two are in contact." I wrote on the board:

$$\text{average acceleration} \quad a = \frac{v_i - v_f}{\Delta t}.$$

"The initial velocity is 2π meters per second and the final velocity is 0 meters per second. But we don't know how long the two are in contact with each other."

"I know how long a ball and bat are in contact in baseball," said Angus.

"You do?" said Marsha dubiously.

"Yup. I play for the Roberts baseball team. It's 7 milliseconds. Our coach likes to say, 'You got 7 milliseconds to do what you're gonna do.'"

"Okay," I said. "That sounds like a reasonable number. If it works for baseball, it probably works for bashing a zombie's head. So let's say 7 milliseconds." I picked up the calculator on Jessie's desk.

"Then the force is

$$F = m\left(\frac{v_i - v_f}{\Delta t}\right) = 1(\frac{2\pi - 0}{0.007}) \approx 900 \text{ newtons}$$

where one newton is one kilogram-meter per second squared."

(II. Force continued on p. 159)

"Jessie, do you know how much force it takes to crack a human skull?" I asked.

"You have a slightly warped view of what I do as a biology professor."

"I know it," said Angus.

"You do?" said Marsha.

"The coach was trying to convince Seth Gamsky not to be afraid of the ball when he's up at bat. He was always scared he was gonna get beaned, and so he always leaned too far away from the ball. The coach explained that it takes 10,000 newtons to fracture a skull. A baseball weighs 5 ounces.

So even if it's going 90 miles an hour, I don't remember the numbers, but it doesn't have enough force to crack a skull."

"I can help you with that," said Gunderson. He took the calculator from me.

"Converting to metric, a ball going 90 miles per hour is going 40.2 meters per second. And 5 ounces converts to 0.145 kilograms. So if it comes into contact with the head for .007 seconds, then the force is given by

$$F = \frac{.145 \times 40.2}{.007} = 828 \quad \text{kg.m./sec.}^2$$

So the ball only creates about 800 newtons. And to crack a skull takes more like 10,000 newtons."

"Yeah," said Angus. "That was supposed to make Seth feel better when he was up at bat."

"Funny coincidence," said Gunderson. "828 newtons is pretty close to the average force you get by swinging your panty hose, Williams. Doesn't sound like you'll be cracking any zombie heads with your toy after all."

"Did it help Seth to know that?" asked Marsha.

"Yeah, he got right in there after that. Unfortunately, he did end up getting beaned."

"Was he okay?" asked Jessie.

"Well, his skull didn't crack, just like the coach said. But it did knock him unconscious."

"I guess that's really all we need, isn't it?" I said. "We just need to be able to knock a zombie unconscious."

"Yeah," said Angus.

"Knocking someone unconscious is all about a sudden sharp blow to the head," said Jessie. "When the head jerks, the brain gets knocked around inside the cranial cavity and it causes a short circuit."

"Works for me," said Angus.

36

At that moment, my cell phone rang. Before I could answer it, Jessie's desk phone also rang. At the same instant, Angus's cell phone, which was in his front jeans pocket, also rang. The three of us simultaneously answered our phones. We all heard the same message.

"This is the Roberts College Campus Emergency Phone System. You are receiving this call because there is a campus emergency. Please proceed at once to the nearest campus emergency shelter. You will receive further instructions there. To repeat, please proceed with haste to the nearest emergency shelter. You will receive further instructions there. This is a message from the Roberts College Campus Emergency Phone System."

"Where the hell is the nearest campus emergency shelter?" asked Angus.

"I have never even heard of a campus emergency shelter," said Jessie. "And I've been a faculty member here for twelve years."

"Well, glad to know the college is on top of the situation," said Gunderson.

Hoyle's head dropped down and he slumped in his chair. I quietly worked my way around until I was directly behind him and then reached down to get hold of the butt of his gun. I had it halfway out of the holster when he grabbed my hand.

"What the hell are you doing?" he yelled, as we struggled over the gun.

Jessie took one step forward, wound up, and swung her loaded stocking. It caught Hoyle in the jaw and snapped his head back.

I grabbed the gun and pointed it at him.

He looked stunned, and then glared at Jessie. "That is going to land you in prison, lady."

"I apologize, chief, but honestly, right now, that's the least of my worries."

Hoyle rubbed his jaw. A large black bruise was beginning to appear. But he remained seated, and quickly dropped back into his previous stupor.

"We need to tie him up," said Jessie in a whisper.

"You could have used my stockings for that," said Marsha, "but now I'm out."

I pulled an extension cord out of the wall socket. "We'll use this." Then I spied the cord to Jessie's backup drive for her computer. "And this," I said.

Hoyle was too far gone to resist, and we tied him carefully to the chair. He was almost docile, and the few things he did mumble were incoherent.

"Okay," Jessie said. "Now we need a plan. We can't get out the door. There are too many zombies out there. It's too risky."

"What about the window?" suggested Angus.

I could see that there was a handle on the bottom half of the window that allowed you to tilt the bottom pane in at least a bit, but only so far before it banged into the radiator. I opened it and pulled it as wide as it would go.

"Doesn't look too big," said Angus.

"Yup, only someone relatively small is going to fit through there."

We looked at Marsha and Jessie, the only viable candidates.

"I'm not going to leave the rest of you behind," said Marsha.

"Well," I said, "the idea would be to not leave us behind. Someone could go out that window, and use the ledge to get to the window of the next office. Then, he or she could go in that window and come out that office door."

"Yes? And then what?" asked Marsha dubiously.

"And then he or she could run like hell and attract all the zombies away from this door. Then the rest of us could escape, too."

"Escape to where?" asked Gunderson.

"I don't know about you, but I have to get home. I have two kids on their own there, and I need to get there as soon as possible."

"I'll go with you," said Jessie.

"Well, I'm not going back to my apartment by myself," said Marsha. "I'll come, too. It's probably safer if we stick together."

"Personally, I think I have a better shot by myself," said Gunderson. "I'll stick with you until we get out of the building, and then I'm going it on my own."

Angus pointed at the car smashed in the middle of the quad.

"We have to help Dr. Ortiz," he said.

"Come on," said Gunderson. "In the shape he's in, what's the point?"

"What do you mean by that?" asked Angus.

"Look, we know how fast the zombies can move. A yard a second. Since there are 1760 yards in a mile, that means it takes a zombie 1760 seconds to go a mile. Divide by 60 and that's around 30 minutes a mile. So that means zombies move at 2 miles per hour. That doesn't sound like much, but I'm guessing they can sustain that for a long time. Anybody that moves slower than that is zombie meat. There's no point in risking our lives for slow people. Slow people are eventually going to get caught. Ortiz is not going to be capable of moving that fast."

"It's not just your speed that matters," said Jessie. "Smarts, strength, common sense. They're all important."

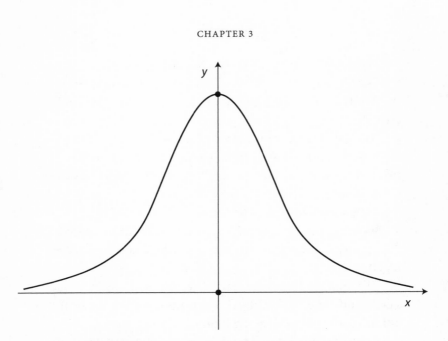

Figure 3.1: The standard normal distribution (bell curve).

"Yes, but regardless, if someone isn't fast enough, they're dead. Just think about the various speeds that people can run. And I don't mean the 100-yard dash. I mean how fast they can run for a sustained amount of time. Those speeds are normally distributed around the mean. And the average time for someone to run a mile, excluding small children and bedridden invalids, is about 12 minutes. So the mean speed is about 5 miles per hour."

"What's normally distributed mean?" asked Angus.

"Haven't you had any statistics? You know the normal curve?" asked Gunderson. "Is this picture ringing any bells?"

He drew Figure 3.1 on the board.

"That does look like a bell," said Angus.

"That's the standard normal distribution. It's given by

$$f(x) = \frac{1}{\sqrt{2\pi}} e^{\frac{-x^2}{2}}.\text{''}$$

"That's a pretty complicated equation just to get a bell," said Angus.

"It's not that complicated," said Gunderson. "Look. The curve is symmetric about the y-axis, since $f(x) = f(-x)$. That means it is its own reflection over the y-axis."

"I see that," said Angus.

"And its derivative is $f'(x) = \frac{-x}{\sqrt{2\pi}} e^{\frac{-x^2}{2}}$," continued Gunderson. "That equals 0 only when $x = 0$. So it only has the one critical point."

"What's a critical point?" asked Marsha.

"A critical point is an x-value where the derivative equals 0," said Angus. "Then the tangent line's horizontal. That second part in the equation, the $e^{\frac{-x^2}{2}}$, will never equal 0."

"That's right, Angus," I said. Gunderson ignored me.

"Clearly," he continued, "as x gets large, either positive large or negative large, the function $f(x) = \frac{1}{\sqrt{2\pi}} e^{-\frac{x^2}{2}} = \frac{1}{\sqrt{2\pi} e^{\frac{x^2}{2}}}$ shrinks to 0. So it peaks at $x = 0$, when $f(x) = \frac{1}{\sqrt{2\pi}}$, and it shrinks to 0, for any large positive or negative x. So it looks like a bell."

"But why is the $\frac{1}{\sqrt{2\pi}}$ in the front? That just makes it look ugly," said Angus.

"That's so that the total area under the curve is exactly equal to 1," replied Gunderson.

"That area can't be equal to 1," said Angus. "It keeps going to the left and to the right. The area is infinite."

"That's right," said Marsha. She and Angus bumped knuckles.

Gunderson looked at me and gave a helpless shrug.

"You actually can have a region with infinitely long tails," I said, "but with the total area still finite."

(III. The normal distribution continued on p. 162)

"Okay ...," said Angus.

"Okay," said Gunderson. "So now we want to be able to center the bell at somewhere other than $x = 0$. So we replace x in the equation of the curve with $x - \mu$. That shifts the peak to $x = \mu$ instead of $x = 0$."

"Okay ..."

"Then we add in a factor of $2\sigma^2$ to the denominator of the exponent in the equation. That allows us to widen or thin out the bell shape by changing σ. So σ determines the width of the bell shape. That gives us the normal distribution

$$f(x) = \frac{1}{\sigma\sqrt{2\pi}}e^{\frac{-(x-\mu)^2}{2\sigma^2}}.$$

The parameter μ is the mean, and σ is the standard deviation." He drew Figure 3.2.

"Zombies are deviants," said Marsha.

"I said deviation, not deviant," said Gunderson.

"Well, maybe one is the noun and one is the adjective," said Marsha, obviously pleased with herself.

Gunderson shook his head in disgust. "I don't even know what to say to that," he said.

"The equation is kind of messy," said Angus. "Couldn't you use a simpler one?"

"First of all, relatively speaking, it's not messy," said Gunderson. "And second of all, everything that is there is there for a reason."

"If you say so," said Angus.

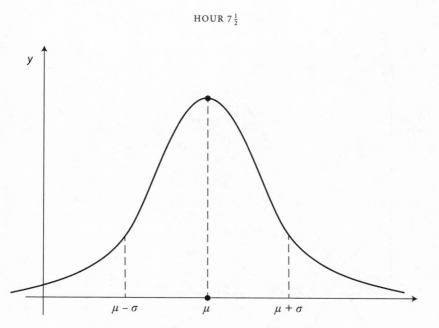

Figure 3.2: The general normal distribution.

"I do. So now look at various people's speeds," continued Gunderson. "The mean is $\mu = 5$ mph. I don't know exactly what the standard deviation σ is, but it's probably around $\sigma = 3$. That means that we expect that the percent of people who can run between 2 mph, which is essentially a walk, and 8 mph, is 68.3%."

"Where did that percentage come from?" asked Angus. "Thin air?"

"If you knew some statistics," said Gunderson peevishly, "you would know that's just a fact about the bell curve. The percent of the total area under the curve that is within one standard deviation to either side of the mean is always 68.3%. Within two standard deviations is 95.5% and 3 standard deviations is 99.7%. It's called the 68-95-99.7 Rule."

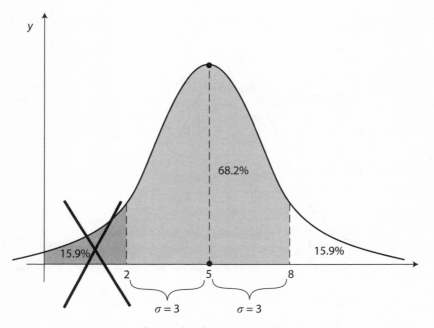

Figure 3.3: The people who run slower than the zombies.

"Not a very catchy name," said Angus.

"It doesn't need to be catchy. It just needs to work."

"So what's your point?" asked Angus.

"My point is that this means 31.7% do not run between those speeds. So we expect half of those to run slower and half of those to run faster. In other words, say about 16% run slower than the zombies move. So we can essentially eliminate them from consideration. They are dead." He crossed out the left shaded region as in Figure 3.3.

"So what does any of that have to do with anything?" asked Marsha.

"How fast can you move in those heels?" asked Gunderson.

Marsha looked down at her feet. "Not very fast," she said slowly.

"So you're over here," said Gunderson, pointing at the crossed-out region.

"I have sneakers in my office," said Marsha.

"Yeah? Good luck getting those," said Gunderson.

"What about Ortiz?" asked Angus.

"In the shape he's in after ramming that tree, Ortiz is over here, too," continued Gunderson. "He's a dead man, even if we get him out of the car. There's no point risking our lives to try to save him."

"You know, Gunderson," I said, "the human race hasn't survived for over 200,000 years by leaving everyone to fend for themselves. We've succeeded by cooperating with one another. That's what we need to do. If Ortiz isn't capable of defending himself, then it's incumbent upon us to help defend him."

"So we will go rescue him?" asked Angus hopefully.

"Before we consider anything like that, we have to figure out how we're going to get out of this office," I said.

"I'll go out the window," said Jessie.

"No, I'll go," said Marsha.

"Given the heels, Marsha, I think Jessie's elected," I said, knowing that if Jessie didn't make it, I would regret this decision. I turned to Jessie.

"Plan is for you to go out the window, along the ledge, in the next window, and then out the office door, making enough noise to attract the zombies away from our door. Then you go out the stairwell and up the stairs to the Psych Department. Once the zombies follow you down the hall, we head the other direction and up the other stairwell. We'll meet you in the Psych seminar room on the third floor."

"What if the window to the next office is locked?" asked Jessie.

"That's what this is for," I said, holding up the stocking weighed down by the crystal paperweight.

"Okay," said Jessie. "Let's do it."

She leaned down and, bracing herself with her hands on my shoulders, put one foot out the window onto the ledge. As I helped to support her upper body, she carefully placed her second foot on the ledge and scooted down to squeeze her hips through the opening. Then she worked her chest through until she was standing outside the window on the ledge. It was a good two feet wide, making it a relatively safe place to stand. Through the open window, I handed her both the weighted stocking and the gun.

"Wish me luck," she said as she tucked the gun into the waist of her jeans. She cautiously crawled along the ledge until she reached the next window.

"It's locked," she yelled back. "And this is as far as the ledge goes."

"You'll have to break it," I yelled back.

I could just barely see her because of the angle, but I saw the light glint off the crystal paperweight as it swung back, and then arced toward the window. Then I heard the glass shatter.

So did most of the zombies in the quad, and a large contingent quickly turned from the smashed car and lurched their way toward the Science Center. Jessie reached through the broken glass to unlatch the window. We were all so busy craning to see how she was doing that none of us noticed that Hoyle's eyes were open. I heard the moan come out of him, the same rusty moan I had heard from the zombies outside the door, but it took me a second to realize what it was. I turned around just in time to see Hoyle standing right

46

behind me, still tied to the chair, his jaw chewing vigorously. Marsha screamed as I fell backward onto the desk, Hoyle on top of me, desperately trying to reach my neck with his incisors. I struggled with him, but his determination and bulk had me at a serious disadvantage.

Out of the corner of my eye, I saw movement outside the window. Then came a loud bang, coincident with the shattering of the entire window for the office, and Hoyle fell off me onto the floor, blood gushing out of a large hole in his forehead.

I looked up at the window where I expected to see Jessie still holding the gun, but she was gone.

chapter four

HOUR $7\frac{3}{4}$

"What happened?" I asked. "Where is she?"

"She fell backward off the ledge when the gun went off," said Angus.

"Oh, my God," said Marsha, holding her hand over her mouth.

"Good plan, Williams," said Gunderson.

I jumped to the window, and leaned out to look down. It was only one story, but then again, there were dozens of zombies directly below the window.

"I don't see her," I said desperately, my stomach in a knot. "Where is she?"

"I don't know," said Angus. "She fell right into the middle of them."

"Give me the other stocking," I said to Marsha. I kicked out some of the remaining shards of glass, and then stepped out onto the ledge.

"I'll meet you in the Psych seminar room upstairs," I said. Peering over the ledge, I saw no sign of Jessie in the milling zombies. They all looked up at me hopefully. Edging as quickly as I could along the ledge, I reached the next window and climbed in through the broken glass. It was one of the chemistry labs.

Putting my ear against the inside of the door, I listened for any noises emanating from the other side, but the moans of the zombies were clearly coming from down the hall. Throwing open the door, I sprinted toward the stairwell. The zombies that were crowded outside Jessie's office turned as they heard the door open and immediately lurched down the hall in my direction. The stairwell door was still propped open by the body of the student who had unfortunately opened the stairwell door. As I leaped over her prone form, I realized there was a zombie crouched over her, feeding. I registered the brown clothing of the UPS delivery guy as I swung the mug-in-a-stocking just as he was looking up, catching him a good blow in the temple. His head snapped to the side as I bounded past him down the stairs.

As I leaped down the last few stairs, I spotted Megan standing in the middle of the lobby. She heard me and turned to face me, her head at an even more distressing angle. With enough random attempts, she must have managed to press down on the door handle and unlatch the door to her prison. She seemed almost surprised to see me, if zombies can look startled. But certainly pleased nonetheless. Her white blouse was stained red, and one sleeve of her cardigan had been ripped off.

She came at me, but with lots of room in the lobby, I got a good hard swing in, and with her weakened neck, the blow seemed to do some real damage. She went down hard.

I looked out the lobby doors to the crowd of zombies congregated there, but there was no sign of Jessie. Turning right, I sprinted down the hall to the other door that opened on the science quad. Listening for a moment, I heard no activity outside the door, so I cautiously pushed it open. There was Jessie clinging close to the wall.

"Thank God," she said as she slipped inside. "That door's locked from the outside."

"I thought you were dead," I said as I shut the door behind her. What happened?"

"I never shot a gun before in my life," said Jessie.

"Welcome to the club."

"I saw Hoyle about to take a bite out of you, so I shot him. Did I get him?"

"Oh, yes, you did. I owe you big time."

"Well, I wasn't prepared for the kick of the gun, and I lost my footing on the ledge and tumbled over, into the zombies. Landed right on top of them. Knocked a bunch over like bowling pins. In the confusion, they couldn't figure out who to bite. I crawled out of the mass of arms and legs, and crawled around the corner before they realized I was there. That's where you found me."

"Thank God you're okay," I said as I gave her a hug. She hugged me back and then handed me the gun.

"Somehow I hung onto this."

"Come on. We have to get up to the third floor where we're meeting everyone else. And I'm guessing we'll have company here, unless we get moving."

We walked down to the second stairwell, far from the stairs I had originally descended. As we climbed the stairs, I could hear grisly noises emanating from the second floor, but there was no activity in the stairwell. We got to the third floor, and I was reaching for the fire door handle when Jesse suddenly screamed, "Look out!"

I turned to see the distorted face of Dan Dreyden, a member of the Computer Science Department, who must have been on the stairs above us. Before I could pull the gun out of my belt, he slammed into me, knocking me backward into the wall. The gun skittered across the concrete floor. I miraculously managed to stay on my feet, and I wrapped one hand around his throat, keeping that arm straight to prevent him from biting me. Then I pushed off from the wall and slammed him back against the stairway railing. With my free hand, I grabbed him by the belt and hoisted him up over the bannister. Any thinking human being would have done his best to shift his weight to avoid the inevitable result, but nobody is likely to call a zombie a thinking human being. I watched as Dan, known to his appreciative students as Dr. D, tumbled backward over the bannister.

As I watched him fall, I counted out loud, "One, two."

He hit the tile floor in the basement with a crunch, and lay still.

"Are you okay?" asked Jessie, putting a hand on my arm.

"Yeah," I said, as my heart rate began to slow to normal.

"Why were you counting?" asked Jessie.

"Reflex," I replied. "I always count when something falls."

"Why?"

"It's a way of telling how far it fell."

"What do you mean?"

"You know, the acceleration due to gravity near the surface of the earth is 32 feet per second squared. So that means that you can get the velocity of an object by integrating the constant 32, and you get

$$v(t) = \int 32 \, dt = 32t + C.$$

We can take the $C = 0$, since he started at rest."

"Okay . . ."

51

"Then integrate again to get the position function

$$f(t) = \int v(t)\,dt = \int 32t\,dt = 16t^2 + C.$$

Again we can take $C = 0$, since we assume this is the starting place. So the distance he has fallen is $f(t) = 16t^2$."

"Okay, so answer the question. Why did you count?"

"I was counting the seconds for Dan to fall. It took 2.5 seconds, which means he fell $16(2.5)^2 = 100$ feet. It's just a quick way to tell how far something falls."

"Did anyone ever tell you you're weird?"

"Yup," I replied, smiling. "Plenty of people. Okay, when I open the fire door, we get down to the Psychology seminar room as fast as possible."

I depressed the door handle and then pulled open the door. Jessie and I sprinted down the hall. When we reached the seminar room, I threw open the door and stopped dead in my tracks. There in a pool of blood on the floor was most of the Psych Department, and the chair, Bob Edelman, was crouched over one of them, blood dripping from his chin.

"Sorry to interrupt," I said, as I shut the door quickly. "We'll need to find a different room," I said to Jessie. Then I saw Angus waving to us from a door down the hall.

We sprinted down to it and shut the door behind us. Marsha and Gunderson were already inside. Marsha gave Jessie a big hug. "Oh, I didn't think we would see you again, at least not in human form."

"I have Craig to thank for that," said Jessie.

"He got you in the fix in the first place," said Gunderson.

"I volunteered," said Jessie.

"Okay, folks," I interrupted. "We need a plan."

Jessie raised a hand to get our attention.

"We need to think about the situation. This isn't a zombie movie. This is the real world. Zombies are living creatures. As such, they cannot function without sustenance. They need food to live."

"All living creatures do," I said.

"But I thought they aren't living," said Marsha. "They're dead. The walking dead."

"You can't be dead and alive at the same time," continued Jessie. "Your brain can malfunction, but to be able to walk around, to be able to give chase, to be able to eat, you have to expend energy. And that energy has to come from somewhere."

"That's not the way it works," said Angus. "Zombies don't need energy. They're like the Energizer bunny. They just keep going."

"Angus, that's the movies, not reality," said Jessie. "There is nothing on earth that can keep going without an energy source."

"What about water?" asked Angus. "It keeps running all the time. And there's no energy put into it."

"The sun is the energy source," I said. "It causes the water to evaporate and eventually it condenses again and comes down high in the mountains and recirculates. But there is an energy source."

"But I still say zombies don't need energy," insisted Angus.

"So, Angus, you're saying that zombies are perpetual motion machines," said Gunderson.

"Yes," Angus said hesitatingly.

"So you think we could harness one to a treadmill, and then have some uninfected person stand in front of it, and the zombie would be so crazy to eat that person that they would drive the treadmill without any need for sustenance or energy input, for as long as we needed them to."

"Ummm, I'm not sure."

"We could take over the fitness center in the gym," said Gunderson. "Hook a zombie up to each of the tread-mills, and generate enough electricity to light up the entire campus. Who needs solar? It's a miracle. And we thought zombies were a bad thing." Gunderson laughed.

"Well, maybe not," said Angus, looking down at the ground.

"Even zombies need an energy source," said Jessie. "Right now it's coming from humans. The zombies have an almost limitless supply of food."

"What do you mean?" asked Marsha.

"The human population. Over seven billion humans. That's a lot of food. But as the number of infected increases, the available food source shrinks. Given enough time."

"Why don't they eat each other?" asked Marsha.

I shrugged. "Jessie?"

"Not sure, but it's not that surprising. Praying mantises excepted, very few species eat their own kind. It's hard to successfully procreate if you eat your mates. So evolution frowns on that, with a few notable exceptions. If the virus caused infected individuals to eat each other, the disease would quickly be snuffed out."

"So what does that mean for us?"

"You mean to survive? It means we have to get away from population centers. Places where they have food to eat. The farther we get away, the sparser the zombies for two reasons. One, there are fewer humans that could be turned into zombies and two, there are fewer humans that zombies can eat."

Angus suddenly pointed out the window.

"Look," he said. "There's Dean Collins."

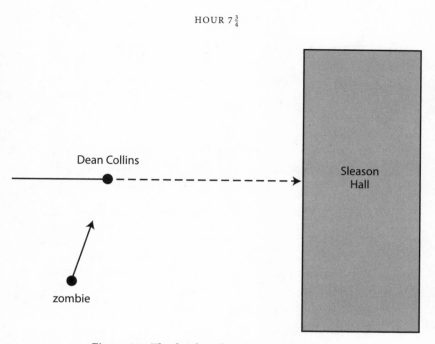

Figure 4.1: The dean's path and the zombie's path.

Collins was the dean of the faculty. He had been tall and agile in his day, a basketball player in college before he went on to get a Ph.D. in classics. But now he was pushing 65, and he had been suffering from hip trouble. We watched as he did his best to evade a zombie that had spotted him from several yards away.

"Classic pursuit problem," I mumbled.

"What does that mean?" asked Angus.

"Watch," I said. "The zombie's goal is to catch the dean and the dean's goal is to evade capture. Dean Collins is headed for Sleason Hall, so he's taking the straight line path to get there. If the zombie had a brain left in its head, it would cut him off, aim for a point on the path that Collins has to take."

I drew the picture in Figure 4.1 with chalk on the blackboard mounted on the wall.

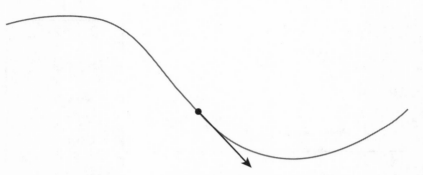

Figure 4.2: The tangent vector to a curve always points in the direction of motion.

"But notice the zombie's tangent vector is pointed straight at the dean at all times, not in front of him."

"Now you've lost me," said Angus.

"Me, too," said Marsha.

Gunderson sighed. "I'll explain it," he said. "If you're moving along a path, your tangent vector is the arrow that points in your direction of motion. Its length is the speed at which you are moving at that instant."

He stood up and drew the picture in Figure 4.2 on the blackboard.

"But the zombie's direction is always changing. So which way should the arrow point?" asked Angus.

"That's the point," said Gunderson. "As the zombie changes direction, his arrow, the tangent vector, points in the direction he is headed in at any given instant. It changes direction as he does."

"What's a vector?" asked Marsha.

"If you don't know about vectors, I'm not going to explain it to you now," said Gunderson.

"A vector's like an arrow," I said. "It has a direction and length. But the nice thing about a vector is that if you keep

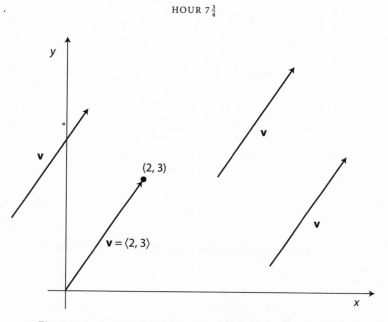

Figure 4.3: A vector has a direction and a length but it can start anywhere.

the direction and length the same, you can start it anywhere. So if the vector is the tangent vector, you start that arrow at the point on the curve for which it is a tangent vector." I drew Figure 4.3 on the blackboard.

"So how do you figure out the tangent vector at any given point on the curve?" asked Angus.

"Well," continued Gunderson, "suppose the original curve of motion is given by a function $\mathbf{p}(t) = \langle f(t), g(t) \rangle$. The $\mathbf{p}$ stands for position function. It's given by two functions. The first, $f(t)$, gives the x-coordinate at time t and the second, $g(t)$, gives the y-coordinate at time t.

"Then the tangent vector is given by taking $\mathbf{p}'(t) = \langle f'(t), g'(t) \rangle$. We just differentiate each of the coordinate functions. The resulting vector, which we call the velocity

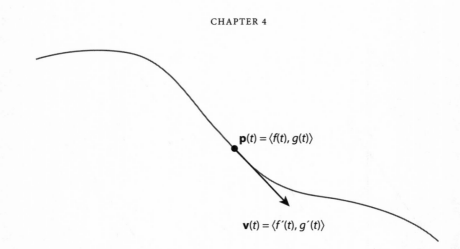

Figure 4.4: The tangent vector is given by $\mathbf{v}(t) = \langle f'(t), g'(t) \rangle$.

vector $\mathbf{v}(t)$, will always be tangent to the curve, and therefore point in the direction of motion at the point on the curve, and its length will always be the speed of the person."

Gunderson drew Figure 4.4 on the blackboard.

(IV. The tangent vector continued on p. 166)

(IV. The tangent vector continued on p. 166)

The dean continued to hobble across the quad, as the zombie continued to head toward him.

"But the dean's tangent vector doesn't change direction," said Marsha.

"That's right," I said. "Because Collins is moving in a straight line, his tangent vector is always pointed in the same direction, straight toward the doors to Sleason."

"But then what was your point?" asked Jessie.

"Well, look at the direction the zombie is headed. His tangent vector is always pointed straight at where the dean is at that instant." I added three instants in time to the first drawing on the board (Fig. 4.5).

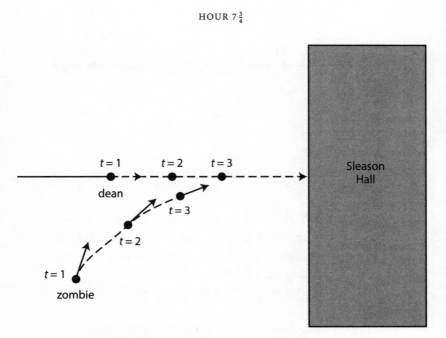

Figure 4.5: The zombie's tangent vector points toward the dean.

"So he ends up following the dean's path rather than cutting him off. He should head to where the dean will be in a few seconds and then he could cut him off."

"So what does the path for the zombie look like?" asked Jessie.

"It's called a radiodrome," I replied.

(V. Pursuit continued on p. 169)

"Usually people describe it in terms of the path a dog would take to chase a rabbit that is running in a straight line at a constant speed. Dogs do the same thing. They head for where the rabbit is at any given instant. The interesting part is that if the dog has the greater speed, he will always

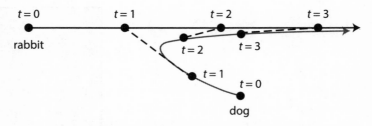

Figure 4.6: Slower dog chasing rabbit that passes nearby.

eventually catch the rabbit, assuming the rabbit doesn't make it to its rabbit hole first. But if the rabbit has the same speed or greater speed than the dog, the dog can never succeed in catching the rabbit."

"Even if the rabbit heads almost straight by the dog?" asked Jessie.

"Yes," I said, drawing Figure 4.6. If the dog isn't smart enough to cut off the rabbit, and just heads to where the rabbit is at that instant, the rabbit always gets by."

"If the rabbit comes really close to the dog, can't the dog just lean over and bite the head off the rabbit?" asked Gunderson, smiling innocently.

"I was assuming an idealized model," I replied, "where the rabbit and the dog are each represented by points. It's true that in the real world, if the rabbit comes close enough to the dog, the dog can catch the rabbit."

"In case you hadn't noticed, Williams, we live in the real world. Idealized models are a waste of time. But maybe you pure mathematicians are unaware of that fact."

"Oscar, be nice," said Jessie, frowning at him. He continued to smirk.

We turned to watch the slow motion pursuit in the courtyard below. As the dean came within a few yards of Sleason, the zombie continued to close the gap.

"The dean has the slower speed," said Angus.

"Yup," I said. "Not good for him."

We watched as the zombie reached out and grabbed the collar of the dean's coat. Collins shrugged out of the jacket and released it, flipping it over the head of the zombie. Then he pulled open the door to Sleason and slipped inside, pulling the door shut behind him, as the zombie struggled to free its head of the jacket.

"Not bad," said Angus.

"All right, Dean Collins," said Marsha. "I knew he could pull it off."

Suddenly, the door to Sleason flew back open and Dean Collins came stumbling back out. Right behind him, a stream of zombies poured out of the building. Colliding with his initial pursuer, Collins tumbled to the walkway, and the zombies immediately fell upon him.

"Too bad," said Angus softly.

"He was a good dean," said Jessie.

"That's not what you said when he turned down your request for a personal leave," said Gunderson.

"Shut up, Oscar," said Jessie, giving me a sidelong glance.

"Personal leave?" I asked.

"Now is not the time," said Jessie. "We need a plan."

"I don't think we should stay here," I said. "We should only stay in rooms with more than one exit, so we can't get trapped so easily."

"What about Ortiz?" asked Angus. "We need to rescue him."

I looked out on the science quad and saw that Ortiz appeared to be the last of the uninfected in the quad. It reminded me of waiting for a bag of popcorn to pop in the microwave. At the end, you get a last pop or two, and then it's done. Ortiz was the last unpopped kernel. Several zombies had returned to Ortiz's car and were slobbering over the windows, but the rest were wandering listlessly around the quad.

"Angus, if we go out there, all those zombies wandering around will come after us."

"Exactly," said Angus. "Just like you said."

chapter five

HOUR 8

It was hard for me to believe I had agreed to the plan, but Raphael Ortiz was someone I had known and respected since my arrival on campus, and I couldn't bring myself to leave him to die. I waited next to Angus and Jessie inside the door through which I had previously let Jessie into the building. I had the chief's revolver in my hand. Up above, on the third floor, Gunderson and Marsha watched from the open window. Angus nodded, and I carefully opened the door, the revolver pointed out. There were no zombies on the other side, so I slipped through the door, concealing myself as well as possible against the building. Angus slipped out, too and then he sprinted toward the bike rack. Several zombies spotted the quick motion and they turned in pursuit.

Angus reached the rack and grabbed the first bike.

"Not that one," Marsha yelled as Angus jerked at it, realizing only then it was chained to the rack. The first zombie was closing fast. I took aim, but doubted my ability to hit anything further than a few feet away. Angus grabbed the second bike and was relieved to see there was no lock. He pulled it from the rack, threw a leg over the crossbar, and started pedaling.

By now, quite a few zombies were in pursuit, but Angus could pedal a lot faster than they could move. He swerved around several that were headed straight for him and then started pedaling around the walkway that encircled the interior of the quad. As he did so, he started yelling.

"Come on you screwed-up assholes," he screamed. "Come and get me."

He swerved around Karen Holm, who reached for him.

"Sorry for the language, Professor Holm, but come and get me," he yelled back at her. A pack of a dozen zombies was now trailing the bike. As Angus came around the far side of the quad, more joined the chase.

Just as predicted, they didn't have the sense to cut across the quad and go to where he would be by the time they got there. They simply headed toward where he was at the instant, making it easy to stay ahead of them. They seemed to be settling into a circular path on a circle of a slightly smaller radius than the one Angus was taking.

"Slow down," yelled Marsha. "You don't want to lap them."

Angus now had all of the zombies in the quad lurching after him, some a few yards behind and some others quite a bit farther behind. But all were essentially in a pack that was following a circle inside his own circular path. As he passed my hiding place, he motioned for me to go. I waited until the parade of zombies passed, and then crept across the quad to Ortiz's car. I knew that any fast motion might trigger a reaction, so I moved slowly, crouching the entire time. When I reached the car, I could see Ortiz in the passenger compartment. His eyes were open, but there was blood dripping from a large cut on his forehead. I tried the door but it wouldn't open.

Tapping lightly on the window, I said quietly, "Unlock the door, Raphael." He turned to stare at me with an uncomprehending look.

"Come on, Raphael. We don't have a lot of time. Unlock the door!" Out of the corner of my eye, I could see Angus coming around the quad again. It appeared that the mob following him had grown.

"How do I know you're not one of them?" asked Ortiz.

"Do you really think they know your name?" I responded. "Open the goddamn door. I'm trying to rescue you."

Ortiz seemed to finally grasp the situation, and he leaned over and hit the button to unlock the door. I pulled on the handle and the door swung open.

"Can you walk?" I asked.

"I don't know," said Ortiz. "What's going on? I don't understand any of this."

"Just get out of the car."

He went to climb out, but his seat belt jerked him back into the seat. I saw Angus and the zombies approaching.

"Get out of there!" yelled Angus.

I reached across Ortiz and unlatched the seat belt. Then I pulled him from the car. I could see several of the zombies peeling off from the pack and making a beeline for the car. I grabbed Ortiz's arm and swung it over my shoulder.

"Time to get a move on," I said, as I half dragged him across the quad.

Angus continued to yell at the top of his lungs. "Come on, you stupid evolutionary throwbacks. Even if you used to be smart, you aren't anymore."

Ortiz and I reached the Science Center and I banged on the door.

"Open up," I yelled.

Jessie pushed the door open, and we fell through as several zombies arrived. Jessie pulled the door shut. I lowered Ortiz to the floor and helped him prop himself up against the wall.

I could hear Angus on the other side of the door still yelling to attract the zombies in his direction.

"Come on, you demented faculty. You used to be so much smarter than me. But not anymore. Look who's leading the parade now."

"Angus, they're safe," yelled Marsha from her window. "Get in the building."

I couldn't hear Angus's response but the banging noises outside the door had stopped so I cracked it open. I could see Angus at the far side of the quad. All of the zombies were again following him in a pack. He was laughing when he spotted me.

"Look at me," he yelled. "I'm the pied piper of Westbridge." It was at the instant when he was looking back at me rather than ahead that the bike ran into the curb and he flipped over the handlebars.

"Oh, crap," I said. Angus landed in a heap. The horde of zombies closed on him. I knew I couldn't get there in time. Pulling the gun from my pocket, I aimed in the general direction of the mob and fired several shots.

The sound of the gun was enough to attract the attention of a few. I started screaming. "Hey, over here, over here!" But they weren't about to give up on the meal right in front of them.

Angus stood wobbly. "I'm okay," he yelled, waving to me.

"Look behind you," I yelled back, motioning frantically.

He turned just in time to see the first zombie reaching for him. Leaping backward, over the bicycle, he then lifted it up and heaved it at the zombies. Several fell over, tangled in the bike. Then he turned, and sprinted toward us across the quad. There were now lots of zombies between him and us. Here was the true test of the theory. He was faster than they were and they always headed toward where he was. So as long as he didn't get within an arm's reach of them, he could get by before they cut him off. But upwards of thirty zombies were in pursuit.

"Move it, Angus!" I yelled. I aimed the gun toward the mob of zombies behind him, but didn't dare take a shot for fear of hitting him.

There was a row of zombies between him and us, moving toward him on either side of a picnic table. Angus never slowed. He just hit the bench with one foot and the top of the table with the other and then launched himself into the air, sailing over the head of one zombie who futilely tried to grab at him. Angus rolled on landing and then was back up on his feet, sprinting hard.

Just as he was closing on the Science Center, with a clear path between him and the door I was holding open, three zombies rounded the corner of the building, the first grabbing my left arm and pulling it toward its gnashing teeth. Without even thinking about it, I lifted the gun and fired into its face, blowing the back of its head off. It fell instantly. The other two tumbled over the body as they reached for me. I pushed Jessie back into the building just as Angus dove through the door. I pulled it shut as the zombies scrabbled at us.

"Holy crap," said Angus, collapsed on the floor breathing heavily. "I haven't run that hard since I quit ultimate frisbee."

"Angus," said Jessie. "You were magnificent out there." He smiled at her as he pulled himself to a sitting position.

"Thanks, Professor Sullivan. The plan worked, didn't it? Hi, Professor Ortiz."

Ortiz looked at Angus for a moment and then said, "Angus? What are you doing here?"

"He just saved you," I said. "It was his plan."

"What plan?" asked Ortiz.

"He rode a bike in a circle and got all the zombies to follow him," said Jessie.

"The circular pursuit problem goes a long way back," I added.

"Yeah?" said Angus, still breathing hard. "How long?"

"Actually, it dates back to at least 1748."

"That's over 250 years ago," said Angus, visibly pleased. "Was there even calculus then?"

"Oh, yeah. Calculus had been around almost 100 years by then."

"Why were they interested in circular pursuit problems? There weren't zombies then."

"No, there weren't. The problem was first posed in terms of a spider trying to catch a fly walking along the edge of a semicircular pane of glass."

"Really? And somebody cared about that?"

"Angus," said Jessie, "they didn't have TV and YouTube. Entertainment was hard to come by."

"That's true," I added. "There was a lot of general interest in recreational math back then. The problem first appeared in a British journal called the *Ladies' Diary*. Not exactly where you'd expect to find math problems. But they published a variety of math puzzlers." Ortiz seemed to be listening, which I took to be a good sign.

"Eventually the problem was rephrased in terms of a duck swimming around the edge of a circular pond and a dog swimming after it."

"Oh," said Angus. "A classic dog-and-duck problem." He was grinning.

I smiled. "If you say so. But it turns out that even though you can find the differential equations that need to be satisfied, you can't solve them analytically."

(VI. Circle pursuit continued on p. 173)

"What does that mean when you say you can't solve them analytically?" asked Angus.

"It means you can't write down a solution in terms of the usual functions we work with, polynomials, trig functions, radicals, et cetera."

"Then how do you figure out what'll happen?"

"Well, you take the differential equations and solve them numerically. The computer draws the resulting path for you."

"So why did the zombies follow me in a circle that was smaller than the circular path I was following?"

"That's interesting. When you do computer simulations, always assuming that you pedal faster than the speed of the zombie who is chasing you, and still assuming that the zombie is always headed straight for you, then no matter where the zombie starts, it eventually travels in a circle." I pulled a piece of paper out of my back pocket, and drew Figure 5.1.

"So here, where the zombie starts near the right side of your path, it eventually settles into a path that is closer and closer to a circle. That circle is called the *limit cycle*."

"I thought the zombie's tangent vector was supposed to always point at me."

"It does. See, look." I drew Figure 5.2 on the paper.

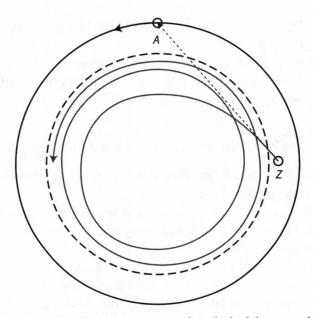

Figure 5.1: Zombie pursuing Angus, who rides his bike in a circle.

"Once the zombie gets to a position like this, then it just follows you around. See its tangent vector is always pointed at you, and the distance between you and the zombie never changes. The size of the circle that the zombie follows is determined by the relative speed of the zombie to you."

"What do you mean?"

"Say the zombie moves at half your speed. Once the zombie settles into what is essentially the limit cycle circle, both you and the zombie travel around your respective circles in the same amount of time, call it t_0. But you travel the circumference of your circle, which is $2\pi R$, where R is the radius of your circle, while the zombie travels $2\pi r$, where r is the radius of its circle. So your speed is $\frac{2\pi R}{t_0}$, and its speed

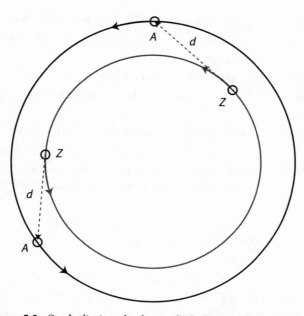

Figure 5.2: On the limit cycle, the zombie's distance to Angus never changes.

is $\frac{2\pi r}{t_0}$. If it travels half as fast as you, then

$$\frac{1}{2} = \frac{\text{zombie's speed}}{\text{your speed}} = \frac{\frac{2\pi r}{t_0}}{\frac{2\pi R}{t_0}} = \frac{r}{R},$$

"So we get $r = \frac{R}{2}$. The radius of the zombie's circle is exactly half the radius of your circle. And, in general, the radius of the zombie's circle will be the fraction of the radius of your circle corresponding to the fraction of your speed that is its speed."

"Then the zombie will never catch me, no matter where it starts. And since the zombies all seem to move at the same speed, it doesn't matter where they start. They all end up with the same limit cycle. Which is why they were all following me on that same circle," said Angus, pleased by his own comprehension.

"Exactly," I said. "And it looked like the circle they were on had a radius of $\frac{3}{4}$ of your circle. So their speed was $\frac{3}{4}$ of your speed."

"And what would've happened if I had slowed down?"

"Then the radius of their circle would have grown. It's radius would shift to be the fraction of the radius of your circle that their speed was as a fraction of your speed."

"But even though they came from different directions, after a while they were all bunched up on the circular path."

"Exactly," I said. "They can start from anywhere, but since they all move at the same speed, they all end up on the same radius circle. And when you were at a point on the outer circle, there was only one point on the inner circle that was behind you and that had its tangent vector pointed at you, so all the zombies ended up in a pack at that point."

"Too cool!" said Angus. "I want to try it again."

Jessie and I traded smiles. "Maybe some other time, Angus," I said.

Suddenly a moan echoed down the hall.

"Time to get moving," I said. "We need to get up to Marsha and Gunderson on the third floor." I helped Ortiz to his feet.

"Crap," said Angus. Down the hall, a lone zombie appeared, shuffling in our direction. Once again, it was Megan, head at a disturbing angle.

"She's determined," I said. "That's what I always liked about her. Wonder if that's a trait that survived infection."

"You want to shoot her?" asked Angus.

"Not really. And anyway, I think I'm down to one bullet. Better save it."

We moved down the hall as best we could, with Megan trailing behind, her arms outstretched, appealing for us to wait up.

"Down here," said Jessie, as she turned down the only available stairs, which led to the basement.

We followed her, with Angus and me supporting Ortiz on each side. At the bottom of the stairs, we threw open the fire door.

"We need to hole up somewhere," I said as we stumbled down the corridor. "Here." I pointed to a door that led to a large storage area where I had kept all my books when I was on leave. The science departments used it to store lab supplies and equipment.

I pulled open the door, and stopped dead at the sight before me. Over a hundred faces turned toward us in terror. A student at the front swung a shovel at my head. I ducked under it.

"What the hell?" I yelled. "We're not infected."

The kid lowered the shovel.

"Sorry, Professor Williams. I didn't know. My job is to protect this door."

I recognized the student from my linear algebra class a couple of years ago. Not the strongest academically, but responsible at the very least. I was pretty sure his name was Todd. The nearest students pushed back against the others to keep their distance from us.

"It's okay. They're human," said Todd to the crowd. "Quick, come inside." Jessie, Angus, Raphael, and I entered through the doorway, and Todd pulled the door shut behind us. I recognized various students in the crowd, including Thora, the multiply pierced girl from calculus. She gave

73

me an almost imperceptible nod. I led Raphael over to the wall, and helped him to sit down with his back propped up against it.

A woman pushed her way through the crowd. I recognized Sylvia Blumenthal, director of the Counseling Center. She was dressed as usual, in a flowing multicolored dress with a large artistic belt buckle, a big chunky necklace, and several filmy scarves. Her long, free-flowing grey hair hung down her back over the scarves. She reached out a reassuring hand and patted my shoulder.

"Craig, I'm so glad you made it here safely."

"What's going on here?" I asked.

"This is one of the Campus Emergency Shelters. Didn't you get the message on your cell phone?"

"Yeah, I got it. I just didn't know where they were."

"There are four, but this one is the biggest. I'm assigned to this one in cases of emergency."

"You're assigned to it? What's your job?"

"I deal with students who have been traumatized in an emergency situation. And believe me, some of these students are definitely traumatized."

"I'm sure that's true, although their emotional state seems the least of our concerns right now."

"Craig," she said, "I'm troubled to hear you say that. How can a student's emotional state not be an important concern? Some of these students have seen horrific events. They will need counseling for years to come."

"That's assuming they survive beyond today," I said, "which as far as I can tell is a big assumption."

"Keep your voice down," said Sylvia, her concerned look fading to a frown. "You're scaring the students. There's no need for that. This is a particularly safe space."

"How so?" asked Jessie.

"There are only two points of egress, the door you just came through and the door at the other end. There are no windows, and we have all the supplies we could need. At least until the rescue team shows up."

"What kind of supplies?" I asked.

"Water here," she said, pointing to a spigot in the wall from which a student was filling her water bottle, "and plenty of freeze-dried food."

"The college stores freeze-dried food in here?"

"It's actually from the Geology Department. They use it when they're out in the field."

"You realize, Sylvia, that at some point, the water is going to stop working. And that will happen well before any rescue squad shows up."

"You don't know that."

"Based on our experiences so far, I think it's a safe bet. And there are several problems with your setup here. Number one, you're blind. You can't see anyone coming. You're just sitting here, waiting for trouble. We just walked in."

"We had guards outside each door, but I'm not sure where they went. Maybe to get help. And infected people cannot open doors."

"Sylvia, I have already seen infected people who opened doors. They just have to try long enough. And number two, there are way too many people crowded in here. If there's trouble, it will cause a stampede. Smaller groups in separate spaces would make a lot more sense. That's a much safer scenario."

"Craig, I know you mean well, but these procedures for emergencies were worked out by experts. The college paid a consulting firm to come in and assess our risk in an emergency. They came up with these procedures. If we don't follow them, we could be culpable."

"You're worried about culpability?"

"Of course I am. That's part of my job description. And speaking of which, why do you have a gun?" I looked down at the revolver gripped in my right hand.

"Strange question," I replied. "We've been using it to survive."

"You haven't shot anyone, have you?" asked Sylvia, a note of incredulity in her voice.

"Not any humans," I replied. "But I did shoot one zombie and Jessie shot another."

"I shot Hoyle," said Jessie.

"The police chief?" said Sylvia. "How could you?"

"He was infected," said Jessie matter-of-factly.

"So what?" said Sylvia. "They're people. They're sick people, but people nonetheless. You couldn't have shot them."

"Well, we did," I said, shrugging.

"There will be repercussions," said Sylvia severely. "You can't just kill people."

"Sorry, Sylvia, but we're just trying to stay alive."

Suddenly, at the far end of the storage area, we could hear banging, followed by what distinctly sounded like a zombie moan. People started surging in our direction. Screams began echoing through the room.

"They're in, they're in," someone wailed over the cacophony of noise. People started pushing toward our end, squeezing others against the wall.

"Oh, dear," said Sylvia. Todd threw the door at our end open and tried to funnel people out. But too many were trying to squeeze out at once.

76

I pulled Ortiz to his feet, and then pulled Jessie to the wall before she was trampled. Bodies surged around us. In the confusion, there was no way to tell who was infected and who was not. Blood spurted from various directions as horrific screams mixed with the moans of the zombies. Each time someone pushed up against us, I fended them off, pushing them back into the crowd, not knowing if they were human or zombie. Suddenly, Thora was shoved up against us. She reached out a hand.

"Professor Williams, please help me," she pleaded. No attitude whatsoever. I grabbed at her hand and we touched fingertips for an instant, but she was pulled away before I could get a hold, and she disappeared into the swirling mass. In the melee, several students tumbled down in front of us, and a pileup quickly formed. For an instant, an opening to the door cleared in front of the pile. Angus grabbed Ortiz by the arm.

"Time to go," he said.

The four of us managed to squeeze out the door. The corridor was filled with people, some hurt, some confused, and some lurching after the others.

We ran down the hall and turned the corner to the left, only to collide with Sylvia, who had been cowering there.

"What the hell, Sylvia?" I said. She just pointed further up the hall, where a half dozen zombies were coming back toward us.

"Safe place, my ass," I muttered. We had only moments. Neither forward down the hall nor back from where we had just come were viable options.

"Shoot them," whispered Sylvia hoarsely.

"What?" said Jessie.

"Shoot them, goddamn it," screamed Sylvia.

I ignored her, and turned to the one door that was left to us. Praying it wasn't locked, I pushed down on the handle. The door swung out, just as the zombies were upon us. We slipped through but when I tried to pull the door shut, I found several zombie limbs preventing me from doing so. "Keep moving," I yelled. We had just entered the bottom entrance to the Wexler Auditorium, the largest lecture hall on campus. It could easily seat five hundred in its raked seating.

Angus hung onto Ortiz's arm to keep him moving, and the five of us cut across the front of the auditorium by the screen, and then up the far aisle, taking several steps at a time.

As we reached the top of the stairs by the main doors to the room, I turned to check on the progress of the zombies giving chase. I actually laughed.

"It's okay," I said. "We don't need to rush."

"Are you crazy?" said Sylvia.

"Look," I said, pointing down to the bottom of the auditorium.

Classic zombie behavior. Instead of following our path across the front of the auditorium and then up the aisle, which any rational creature would have done, the zombies entering the auditorium behind us were again taking the straight-line path between them and us. This meant trying to climb straight up over the seats. There were upwards of twenty-five zombies trying to figure out how to clamber over the chairs. Doing that climb up the auditorium seating would have been difficult for a human, but given the minimal coordination of a zombie, it was truly laughable. They would lift a foot over a chair and then tumble forward landing in the next row. After scrabbling around on the floor a bit, they would eventually right themselves, and then fall over the next chair. In the meantime, they were bumping into each other, knocking each other over like bowling pins. Half the time they fell backward a row. It was pratfall humor that should have only appealed to the lowest common denominator. But given the circumstances, I couldn't help but find it funny.

"Come on," said Jessie. "This should let us lose them."

We threw open the upper doors to the auditorium, and trotted down the hall to the elevator.

"We need to get to Marsha and Gunderson on the third floor," I said. I hit the button to call the elevator.

"Do you really think it's such a good idea to get in an elevator?" asked Jessie.

"Less chance in here of running into a zombie than in the stairwell," I said. The doors slid open and we all crowded inside. Ortiz leaned against the wall, as I hit the button for 3 and the doors slid shut. The elevator was a freight elevator which rose at an interminably slow rate. I began to doubt the intelligence of this decision as we waited to reach our floor. When the elevator finally came to a stop, I stood with the pistol outstretched as the doors slid slowly open. The corridor was empty. I motioned for the others to follow me and, still with the gun at the ready, we moved down the hall until we reached the room where we had left the others.

"Marsha, Gunderson," I said quietly as I knocked lightly on the door. Marsha opened the door, and we crowded in.

"Where's Gunderson?" I asked.

"He said he was going to the bathroom, but he never came back," said Marsha. "Do you think he's dead?"

"I kind of doubt it," I said.

chapter six

HOUR 9

Ortiz sat in the office chair, a glazed look to his eyes. I suspected he had a concussion, but there wasn't much to do about that now. I called my daughter.

"Ellie, you okay?"

"Yeah, Dad. We're in your bedroom, with the dog. Connor has already eaten all the chips and now he's thirsty. But I don't want him to drink all the Gatorade."

"He can drink water from the faucet in the master bath. And tell him to stop eating the snacks."

"When are you going to be here?"

"As soon as I can honey, as soon as I can. Right now, we're still stuck in the Science Center."

"We've seen some neighbors acting weird. Mr. Burton is just standing outside his house in his undershorts, in the rosebushes."

"Just shut the curtains, honey, and don't worry. I'll be home soon."

"Okay, Dad. Please hurry. Bye."

"I thought you said they didn't eat each other," said Angus to Jessie, pointing out the window at two zombies grappling on the ground beneath us, tearing at each other's clothes.

"They aren't eating each other," said Jessie.

"What? Of course they are. You don't mean that... oh my God."

"Like I said, the only functions that still remain in the zombie brain are the earliest instincts. Sex goes all the way back to the beginning."

"I think I'm going to be sick," said Marsha.

"Can they get pregnant? Have baby zombies?" asked Angus.

"If so, it's unlikely the baby would survive," said Jessie. "Humans need a tremendous amount of care for a very long time to survive to adulthood. Without the higher brain function, the mother wouldn't be able to care for her young."

Marsha took a tissue from a box on the desk and dabbed at the blood on Ortiz's forehead.

"How's your head?" she asked.

Ortiz looked up at her. "It hurts, but not so bad. Just feel a little out of it. So what's going on? What is this? A disease? How fast is it spreading?"

"It's exponential growth," answered Angus.

"Well, initially it is," said Jessie.

"What do you mean, initially?" asked Angus.

"Logistic growth is really a better model," said Jessie.

"What's that mean?" asked Angus.

"It means that initially, when there are plenty of un-infected people around, the zombies have lots of people to infect, so the number grows fast. But eventually, the number of uninfected people drops relative to the number of infected. So they don't have the big pool to draw on anymore."

Jessie drew Figure 6.1 on the whiteboard hanging on the wall.

"What's P_0?" asked Marsha.

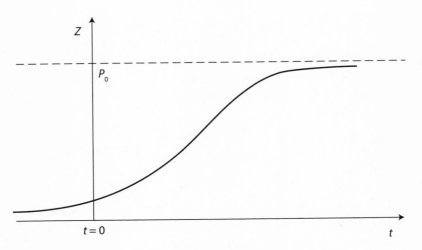

Figure 6.1: Logistic growth of the number of zombies.

"P_0 is the total population before the infection begins. So if we're talking about the United States, it would be just over three hundred million. But let's just talk about the town of Westbridge. Then it's about ten thousand."

"Okay," said Marsha.

Jessie continued. "If we let Z be the number of infected people in town, then the rate at which their number grows will be proportional to the number of infected people, which is Z, times the number of uninfected people, which is $P_0 - Z$. So we get this differential equation." She wrote the following equation on the board.

$$\frac{dZ}{dt} = kZ(P_0 - Z)$$

"Why would the derivative be proportional to $Z(P_0 - Z)$?" asked Angus.

"Think of it this way," said Jessie. "If there are lots of infected people, so Z is big, then the number of infected grows fast, since there are more to infect others. And if the number of uninfected people is large, that's $P_0 - Z$, that also makes it grow fast, as there are lots of candidates for infection. But if either the number of zombies Z or the number of non-zombies $P_0 - Z$ becomes small, then the right-hand side of the equation is small, so the rate slows down, since we either don't have as many to infect others, or we don't have as many to infect."

"But if that's your differential equation, then why does the function look like that elongated S that you drew?"

"Well, look at it this way. We can multiply out the right-hand side to get

$$\frac{dZ}{dt} = k(P_0 Z - Z^2).$$

"When Z is small, Z^2 is a lot smaller than Z. So as an approximation, we can ignore the Z^2 term, and our differential equation looks like $\frac{dZ}{dt} = kP_0 Z$."

"And that gives exponential growth," said Angus.

"Exactly. So for small values of Z, which is during the initial period of the infection, we see exponential growth. That's to the left side of the picture, where you see the curve beginning to rise more steeply.

"But once Z starts getting bigger, and approaches P_0 in size, then $P_0 - Z$ goes to 0, so in the original equation $\frac{dZ}{dt} = kZ(P_0 - Z)$, the right-hand side goes to 0. That will cause $\frac{dZ}{dt}$ to approach 0. But $\frac{dZ}{dt}$ is the slope of the curve, so since it is approaching 0, the curve must be getting flatter and flatter, which is exactly what we see on the right side of the graph."

"That makes a lot of sense," said Angus. "But can you write down an actual function for the logistic growth?"

(VII. Logistic growth continued on p. 175)

"Yup.

$$Z = \frac{P_0 J \, e^{P_0 kt}}{1 + J \, e^{P_0 kt}}$$

is the actual function that solves the differential equation."

"Okay. Then what does that tell us?"

"Well, notice that when t gets large, the 1 in the denominator will be tiny in comparison to the other terms, so Z is approaching $\frac{P_0 J \, e^{P_0 kt}}{J \, e^{P_0 kt}}$, which equals P_0. In other words, wait long enough and everyone eventually becomes infected."

"What do you mean 'everyone'?" asked Sylvia.

"I mean everyone," replied Jessie. "You, me, everyone in this room. Everyone in this building. Everyone in Westbridge becomes a zombie."

"Pretty pessimistic model," I said.

"Yes, it is," said Jessie.

"But what about the J and the k?" asked Angus. "Don't the choice of those influence the model?"

"Yes, they can slow down or speed up the growth. But in all cases, eventually everyone is infected."

"So what?" said Sylvia. "We should just give up? We have no hope of survival?"

"I didn't say that."

"Yes, you did."

"No, I said the model says that eventually we lose. But it's just a model. There are more accurate models out there."

"I hope so," said Angus.

"The government must be aware of what's going on," said Sylvia. "They'll send help. The army will be here soon. We need to sit tight until help arrives. In the meantime, we can be here to support one another."

"Sorry, Sylvia, I can't sit tight," I said. "I have two kids alone at home. I'm going."

"I'll go with you," said Jessie.

"I'll come, too," said Marsha.

"Me, too," said Angus.

"No, it doesn't make sense for a lot of us to go. Better for you all to stay here. Ortiz is in no shape to travel. Jessie and I will go. Angus, give me your cell phone number. I'll call you once we make it."

"You can't leave us here," said Sylvia, resentment edging her voice. "Your first responsibility is to the college. And you have a responsibility to the students. It is your duty to protect them."

"Come on, Sylvia. Are you serious? My first responsibility is to my kids. I'm going to go take care of them and there's nothing you can say that would prevent me from going. And Angus can certainly take care of himself."

"Yeah, I can," added Angus.

Sylvia crossed her arms in a huff but she restrained herself from saying more. I exchanged cell phone numbers with Angus. Marsha handed me the stocking containing the coffee cup, and she gave me a hug.

"Good luck, Craig," she said.

Jessie opened the door and looked up and down the hallway.

"All clear," she said. We slipped out the door.

There were some disturbing noises emanating from the Psychology seminar room, but otherwise, the corridor was quiet. We walked down to the stairwell and peered through the window in the fire door. It appeared clear, so we opened the door and descended the stairs slowly. I held the gun at the ready, and Jessie carried the weighted stocking. We made it down to the ground floor door without incident.

The coast looked clear, so I pulled my keys out of my pocket, and we slipped out the exit and trotted across to the parking lot. Unfortunately, my car was blocked in by several other abandoned cars.

"What do we do?" asked Jessie.

"Where's your car?"

"It's in the parking lot by Prospect."

"Then that's where we're headed. Stay close."

Jessie and I started trotting across campus. But as we crossed the science quad, we came a little too close to the corner of the Brechter Geosciences Building. Suddenly a figure appeared from around the corner and ran straight at us. I tried to shield Jessie, pulling her in the other direction.

"Professor Williams," yelled the person. "Wait!"

We stopped. I recognized Jimmy Langdon, a problem student if there ever was one.

"Professor Williams," he said, when he reached us, out of breath. "I'm really sorry I missed class today. My alarm didn't go off. That happens sometimes. I guess sometimes I forget to set it. But then when my roommate gets back from his 9 o'clock class, he always wakes me up. I mean always. Never fail. But he didn't wake me today. So I'm sorry, but will you take this late homework?"

He waved the paper in front of me.

"Jimmy," I said, "do you really not know what is going on?"

"I'm truly sorry, Professor Williams. I know this is my second late homework, and you made it clear on the first day that you would only take one late homework, but how could I have known my roommate wouldn't wake me, and technically it isn't late because I did finish it on time. I just didn't turn it in on time." He waved the homework in front of my face again but only halfheartedly.

"Jimmy, you have to get out of here, some place safe. You just slept through the beginning of the zombie apocalypse."

His face turned hard. "Okay, if you won't take it, you won't take it. But I am already running a D in your class. This could do me in."

I grabbed him by the shoulders. "Jimmy, you're not listening to me. This is serious. There are people around here who want to kill you."

He shook free of me. "You're the one who's killing me. I could pull out of this tailspin, if you gave me half a chance." He threw the problem set on the ground and started walking away.

"Jimmy, you don't understand."

"I understand just fine," he said over his shoulder. Then he looked up and saw a zombie a distance off.

"That's my roommate," he said. "The one who didn't wake me up. He'll tell you."

He trotted in the direction of the zombie.

"Hey, Marco!" he yelled.

Marco and several other zombies turned in his direction.

"We have to do something," said Jessie.

"Jimmy, come back!" I yelled.

But Jimmy ignored me. As he reached Marco, he must have realized something was wrong, and he stumbled backward, just as two other zombies converged on them. He fell to the ground screaming as they tore at him.

"Come on," I said. "We have to keep moving."

87

Several zombies had heard me yell at Jimmy, and they let out their characteristic moan as they set off after us. This attracted more zombies.

"Craig, this is not looking so good," said Jessie, as zombies headed toward us from a variety of directions. We started sprinting, with a good crowd following behind. Several others appeared in front of us. I pointed to Hopson Hall. "Come on," I said as we headed for the doors. We reached them with seconds to spare. As I pulled the door open, several administrators, including the bursar and the associate provost, came lurching out, for once their incentive unrelated to overspent department budgets. Before I could react, the associate provost knocked Jessie to the ground, and fell upon her.

I swung the weighted stocking and caught him on the temple, knocking him off her, as I simultaneously managed to sidestep the bursar. The crowd that had been pursuing us arrived just in time to join in the fun. I grabbed Jessie by the arm, pulling her to her feet, and we ran down the drive by the construction site for the new library. Unfortunately, the construction had not reached the point where there was any enclosed space that would afford protection from the zombies. In our path was an additional crowd of zombies who had been attracted to the noise and were now lurching toward us. Scanning for anywhere we could hole up, I spotted a lone car sitting in the parking lot.

"Come on," I said to Jessie and we sprinted to it. I pulled on the door handle but it was locked.

"Damn it," I said desperately.

"Craig, we have to do something," said Jessie as she watched the zombies approaching from several directions. I swung the weighted stocking at the driver's side window.

The blow shattered the glass and set the car alarm off. I reached inside and unlocked the door. Jessie gave me a confused look.

"Just get in," I said as I threw open the door. At least thirty zombies were converging on us. She jumped in and I followed, slamming the door behind me. The zombies reached the car. Jessie was pressed against the far door, sitting on a pile of broken glass. I put my back against her and my feet up by the window. When the first zombie reached in, I gave it a good sharp kick to the head, jerking its head back out of the window and embedding a few broken shards of glass in its neck. But it pushed back into the car, unaware of the damage I must have inflicted. Zombies began to crowd around the window trying to reach past it. As I kicked at the first, it did its best to bite my foot. I was grateful I'd gone with the leather running shoes rather than the fabric kind. I pulled back and then with all my force, I kicked it in the head again. Its neck seemed to break, and the head lolled to the side. But amazingly, it continued to try to grab me.

"Unlock the other door," I said to Jessie, "and get ready to run."

I kicked one more time and then yelled, "Now." Jessie threw open the far door of the car and hopped out. I twisted around and jumped out, too, grabbing her arm and heading down hill. The zombies were congregated on the far side of the car, and it took them a moment to understand we were no longer in it. But they quickly absorbed the fact that their prey was in flight and started after us.

It was at this moment that I spotted a blue port-a-potty sitting next to the construction site. We were out of options, so I pulled Jessie toward it.

"This way," I said, urging her on. We reached the port-a-potty, but when I pulled on the door, it was locked from the inside. I slammed my fist against the door.

"Sorry, it's occupied," said a voice from inside.

"Gunderson, is that you?" I said frantically. "Open the door."

"No room in here," replied Gunderson.

"Oscar, open the door," said Jessie.

"Is that you, Jessie? Sorry, but I can't open it."

"Open the goddamn door," I screamed, as I pounded my fists upon it.

"Not gonna happen," he replied.

The zombie horde was a couple of yards away now.

"Look, Craig! There's another port-a-potty behind this one," said Jessie.

We scrambled around the first to the second just as the zombies stumbled over the curb and were reaching for us. I grabbed the door and miraculously, it flipped open. Jessie and I crowded inside as the first zombie craned its neck to take a bite. I gave it a well-placed kick in the solar plexus and it fell back, knocking over several others behind it. Swinging the door shut, I threw the bolt to latch it. The zombies swarmed outside the port-a-potty, banging futilely on the exterior.

"Damn, that was close," I said. The two of us were crowded up against one another, the fecal smell wafting out of the toilet hole mixing with the odor of chemical antiseptics. Jessie was breathing hard. I flipped the lid down over the hole.

"Have a seat," I said. "We may be here a while."

Jessie maneuvered around me and sat down. I leaned against the wall, weaving my feet around hers.

"Nice going," called Gunderson from the other port-a-potty. "You managed to attract every zombie in town. Now we're really stuck."

"Hey, asshole," I said. "You almost got us killed."

"Give me a break," said Gunderson. "The three of us would never fit in one of these things. If I had opened that door, we'd all be dead."

"That's probably true," said Jessie quietly.

"Why are you always defending him?" I asked. "He's a complete dipstick."

Jessie didn't respond. She just looked down at the floor.

"Wait a minute," I said. "Why are you not looking at me? Did you have a relationship with that jerk?"

Jessie shrugged noncommittally.

"You did, didn't you?" I asked incredulously.

"Craig, you pushed me away," she said. "I was hurt."

"And you turned to Gunderson? Gunderson, of all people?"

"It wasn't like that," she replied.

"What was it like?" I asked. I realized how ridiculous this conversation was, the two of us trapped in a port-a-potty surrounded by zombies, the smell of feces and chemicals wafting around us. But I needed to know.

"It started at the New Faculty party at the beginning of the academic year, after you said you didn't want to see me anymore."

"We agreed not to see each other anymore."

"Whatever, Craig. I was hurt. I drank too much wine at the party, and Oscar offered to walk me home. I invited him in, and well, you know."

"I don't know. You had sex? So this was what, a week after we broke up?"

"Something like that."

"You know, I didn't date anyone for a year and a half after that."

"Craig, that wasn't because of me. Your wife had just died."

"How long did you and Gunderson continue?"

"A couple of months. We both didn't want anyone to know. So we just met at his place after that, at odd times."

"What? You mean just sex?"

"I didn't love him or anything, if that's what you mean."

"You make it sound like that makes it okay."

"What? A woman can't have sex for sex's sake?"

"No, I understand that. Just not with Gunderson."

"You don't know him."

"Are you kidding me? You saw what he just did. Give me a break."

"He's insecure. Most of his arrogance is his way of trying to cover up his insecurity."

"And his willingness to watch us get eaten? What's that from? A lack of maternal affection?"

"You know, this is really not the time to be talking about this."

"I'm guessing it was his dream come true."

"What?"

"Having sex with you, and no other commitment. No time put into a relationship. Just sex."

"That wasn't what it was. Why? Is that what you want?"

"I didn't know it was an option."

"With you, it wasn't."

"Why not?"

"I liked you too much."

There was a long pause, broken only by the sound of zombies moaning and banging against the plastic sides of the port-a-potty.

"And how did it end?" I asked.

"I broke it off."

"Why? Wasn't the sex good enough?"

"Craig, why are you pissed? You didn't want to have a relationship. So what do you care who I see?"

"I wasn't ready. That's all."

"Craig. It's been five years!"

"And Gunderson? How could you stand that jerk?"

"I heard that," said Gunderson from the other port-a-potty. "Jessie, are you talking about me?"

"No, Oscar," said Jessie loudly, and then in a whisper she said to me, "Keep your voice down."

"So why did you break up with him?" I asked more quietly.

"He's a little on the selfish side."

"You think?"

I was surprised by the intensity of my jealousy. I guess I had always assumed Jessie wasn't seeing anyone else, perhaps because she still had hopes for us. It was patently unrealistic, given that five years had passed. But somehow I had clung to that. Of course, now that worldview was changing.

Jessie put her hand on my arm.

"Craig, I'm sorry if this upsets you, but I needed to get on with my life. You weren't ready, and I understood that. You know, I lost my best friend when Wendy died. You weren't the only one who needed consolation."

"Have there been others?"

"Craig, it's been five years. Yes, there have been others."

"Anyone serious?"

"There's a guy I've been seeing on and off who teaches anthro at Amherst."

"Yeah?"

"Yes, but I think both of us have been continuing the relationship out of lack of other options. Better to have someone to hang out with than no one."

"Will you continue to see him?"

"Craig, at this point it's no better than 50% odds he's still alive."

"Hey," yelled Gunderson. "Did you tell him about us?"

"Yes, Oscar, I did," replied Jessie.

"Hope there aren't any hard feelings, Williams. You had your chance."

"Go to hell, Gunderson."

"I think we may already be there," said Gunderson.

chapter seven

HOUR 10

"Do you want to sit down?" asked Jessie. "You must be tired standing."

"Wouldn't mind taking a turn," I replied.

As she stood up and we tried to step around each other, I caught a glimpse of red.

"Jessie, what happened to your leg?" I asked.

"What do you mean?"

"There, on the back of your leg. There's a tear in your pants and a massive scratch."

She craned around to see the red scratch running down the back of her leg.

"I don't know," she said. "In all the excitement, I never noticed it."

"It's not deep," I said. "Maybe it's just from the broken glass in that car."

"Yeah, maybe," she said. I reached for it.

"Don't touch it," she said. "It might be infected. Maybe it's from one of the zombies."

"I doubt it," I said. "And even if it is, it could have been from one of them grabbing at you, tearing your pants and scratching you. It doesn't mean that the virus has entered your body."

"Yeah," said Jessie. We were both silent for a moment.

"Umm," said Jessie. "But in case the virus has gotten into my body, I think I should leave the port-a-potty."

"Oh, come on," I said. "That's crazy. You leave this port-a-potty and you are dead . . . or worse. We don't know anything at this point." Jessie looked down at the cut again.

"Even if some virus has gotten into your body," I continued, "it doesn't mean it will spread. We don't even know what causes the virus to flourish once it enters the human body."

"Actually, those kinds of questions have been studied for years. Maybe not for this particular virus. But for viruses in general."

"Yeah?" I said, thinking to distract her from the issue at hand. "What are the relevant factors?"

"Well, assuming it's like most viruses, it's most likely passed through their saliva or blood or other bodily fluids. Like HIV. You get some of their blood or saliva in a cut or in your mouth or eyes, and the virus is in your body."

"So a bite would be the primary way to pass it."

"Yes, but if the zombie eats enough of you, you're not going to become a zombie yourself, because you're dead. Dead dead."

"Okay, but what happens if you're bitten or scratched and not eaten enough to kill you?"

"Well, then the virus can spread through your body. Usually, that means that a virus particle, called a virion, infects a cell. A virion consists of the viral RNA or DNA, encased in a capsid, which is a protein container for the virus. The virion attaches to the cell wall and then inserts the viral DNA or RNA into the cell.

"Typically, each virus attacks a very specific type of cell. They can only latch onto and infect those types. In this case, it looks like they go after brain cells, at the very least, either the neurons or the glial cells."

"But how would they get to the brain cells?" I asked. "If they get into the body through a bite or scratch that's nowhere near the brain?"

"It depends a lot on the virus. A good example is rabies, because it affects the brain. You get a bite from an infected animal, and the virus, which is in the animal's saliva, then enters the peripheral nervous system."

"What's that?"

"That's the nerves and ganglia that are outside the brain and spinal column. It then spreads through those nerve cells until it reaches the brain."

"But I thought there was a blood brain barrier that protects the brain from infection."

"There is, but it protects the brain from infections that come through the blood. It's a collection of so-called endothelial cells with tight junctions between them that separate the circulating blood from the cranial fluid. But there's no barrier between the peripheral nervous system and the brain. The peripheral nervous system feeds into the central nervous system which then feeds directly into the brain. The rabies virus works its way from cell to cell in the nervous system. It's a highway straight to the brain."

"And what happens when it gets there?"

"It causes encephalitis, which is acute inflammation of the brain. And then in most cases, if it hasn't been caught before then, it causes death."

"What do you mean in most cases? We have vaccines against rabies now."

"Yes, and if you're vaccinated before infection or treated shortly after infection, you survive. But once symptoms appear, 99% of the time, you die."

"Bad odds," I said. "But haven't we pretty much eradicated rabies?"

"In the U.S., yes. But worldwide, there are over fifty thousand deaths a year from rabies."

"Really?"

"Yup, especially in India, because they outlawed the killing of dogs. So there're a whole lot of strays."

"Do you think the Z virus is like rabies?"

"In some aspects, it is. The way that rabies has evolved to be so successful at spreading itself is by destroying the usual brain function of infected animals. They become aggressive and bite other animals, which passes the virus on. The Z virus is acting in the same way. It destroys the higher brain function to the extent that only the basest instincts remain, and then the infected individual tries to eat others, thereby passing the virus on."

"So could it be a variant of rabies?"

"Maybe. But there's a fundamental difference. The Z virus acts so quickly. I mean, we're talking fifteen minutes from bite to brain liquification. That's amazingly fast. I'm not sure infection through the nerve cells could act that fast."

"Well, how else could it get to the brain from the bite or scratch?"

"Some viruses hitch rides on healthy cells, like the uninfected white blood cells. Those go all over the body, and the virions go along for the ride. If that's what's happening, then the virus could spread all over the body very quickly. Figure you have about five liters of blood in your body. And it circulates at a rate of five liters per minute, so it really only takes one minute for all the blood to circulate around the body."

"But then wouldn't the blood brain barrier protect the brain?"

"There are certain viruses that are capable of disrupting the blood brain barrier, including HIV and West Nile virus, for example. They cause inflammation of the endothelial cells that make up the blood brain barrier. That allows the virions to slip through."

"And what happens once a viral particle gains access to the brain?"

"Then they have free access to the brain cells. They latch on and insert the viral DNA or RNA. The machinery inside the cell is then hijacked to make copies. Those copies are then disseminated into the brain, causing further spread of the disease. The infected cells become virion factories."

"What happens to the infected cell?"

"It depends on the virus. If it's cytopathic, it eventually destroys the original cell."

"What's cytopathic?"

"That means the virus substantially alters the cell it infects. Usually then the cell doesn't survive long. Since we already know that the virus liquifies much of the brain, it's safe to assume that this virus is highly cytopathic."

"Can we model the growth of the number of virus particles in the body?"

"Yes. There are models. But, first you have to consider how the growth rate of the number of a type of cell works in an uninfected body. We can let b be the number of brain cells, for instance, in a healthy brain. Suppose that the hippocampus, which is where we think new brain cells are manufactured, produces new cells at a constant rate of λ. Notice that it doesn't depend on the number of existing cells, since we're not creating new cells by having all of the existing cells subdivide."

"Then how are new brain cells manufactured?"

"That's not well understood, but believe me, scientists would love to know the answer."

"Yeah, I can see the obvious applications."

"Now suppose that the cells die off at some rate that does depends on the total number of cells b. Let γ be the so-called death rate."

"Okay."

"Then we have a differential equation that governs the number of cells in the brain given by

$$\frac{db}{dt} = \lambda - \gamma b$$

in units of cells per day." She pulled a piece of paper and pen out of her pocket and wrote down the equation.

"That makes sense," I said. "And we can solve that differential equation."

"Right," continued Jessie. "We separate variables and get

$$\frac{db}{\lambda - \gamma b} = dt.$$

Then we integrate both sides:

$$\int \frac{db}{\lambda - \gamma b} = \int dt.$$

"To do the left integral, we can do a u-substitution of the form $u = \lambda - \gamma b$. Then $du = -\gamma\, db$. So $db = \dfrac{du}{-\gamma}$, and we

100

get

$$\int \frac{du}{-\gamma u} = \int dt,$$

$$\frac{-1}{\gamma} \ln u = t + C,$$

$$\frac{-1}{\gamma} \ln (\lambda - \gamma b) = t + C,$$

$$\ln (\lambda - \gamma b) = -\gamma(t + C),$$

$$\lambda - \gamma b = e^{-\gamma(t+C)}.$$

"So with some algebra,

$$\gamma b = \lambda - e^{-\gamma(t+C)}.$$

And then

$$b = \frac{\lambda - e^{-\gamma(t+C)}}{\gamma}."$$

"Okay," I said. "So now what happens long term?"
"We can write this as

$$b = \frac{\lambda - \frac{1}{e^{\gamma(t+C)}}}{\gamma},$$

$$b = \frac{\lambda}{\gamma} - \frac{1}{\gamma e^{\gamma(t+C)}}.$$

Then as t gets large, the term $\frac{1}{\gamma e^{\gamma(t+C)}}$ gets tiny. So b converges to $b_0 = \frac{\lambda}{\gamma}$. That's called the *equilibrium value*. So in your body right now, the number of brain cells is very close to this equilibrium value b_0."

"Okay, I get all that," I said. "So that's what governs the number of brain cells in a healthy brain. Now what happens if some number of virus particles enter your body?"

"Well, normally, we let v be the number of free virus particles in the body. Now, we'll change b to represent only the number of uninfected brain cells instead of all the brain cells. And we'll let i be the number of infected brain cells. The free virus particles will infect the uninfected cells according to how many free virus particles there are and the prevalence of the uninfected cells. In other words, new infections occur at a rate given by βbv, where β is determined by the efficacy of the process."

"What do you mean?"

"Think of βbv as a way to estimate how often virus particles and uninfected cells come into contact with one another and cause further infection. You know, how easy it is for a free particle to find an uninfected cell, and then how easy it is for the virus particle to gain entry into that cell. Those are the factors that determine β."

"Okay," I said. "So the equation for the growth rate in the number of uninfected cells is

$$\frac{db}{dt} = \lambda - \gamma b - \beta bv."$$

"Right," said Jessie. "And then the number of infected cells is growing at a rate of βbv, but these cells don't last that long. They produce some number of virus particles and then they die. So the rate at which the number grows is actually

$$\frac{di}{dt} = \beta bv - \alpha i$$

where α depends on how long an infected cell survives. It turns out that the average lifetime of an infected cell is then $1/\alpha$. In fact, any time you have exponential decay with decay constant α, whether it's the number of people or virus cells or regular cells or even radioactive particles, the average life span is $1/\alpha$."

(VIII. Average life span continued on p. 179)

"Okay. And then what about the virus particles?"

"Well, the greater the number of infected cells, the quicker the number of virus particles goes up. So we expect that the rate of growth of v will be proportional to i. But it's also true that virus particles eventually die after some time interval. So there is a loss in the rate of growth of the number of particles proportional to the number of particles. Then we expect the rate of change of the number of virus particles to be:

$$\frac{dv}{dt} = \kappa i - \mu v.$$

The κ and the μ are constants that depend on the particular type of virus."

"So?"

"So we have three differential equations in the three variables b, i, and v.

$$\frac{db}{dt} = \lambda - \gamma b - \beta b v,$$

$$\frac{di}{dt} = \beta b v - \alpha i,$$

$$\frac{dv}{dt} = \kappa i - \mu v."$$

"Great! So this tells us how these three variables are related," I said.

"Well, yes and no."

"What do you mean?"

103

"This is a system of three nonlinear differential equations," said Jessie. "We can't get an analytic solution."

"Oh yeah," I said. "The two bv terms are products of two variables, making the equations nonlinear. And nonlinear equations are a bear to solve. Too bad."

"But we can still get an idea of what is going on," said Jessie.

"Yeah?"

"Yeah. Before infection, we have no virus particles, so $v = 0$. And, there are no infected cells, so $i = 0$. The number of uninfected cells is at the equilibrium value $b = \lambda/\gamma$. Then, suppose that at time $t = 0$, the infection occurs with the entry into the body of $v = v_0$ virus particles. The critical quantity that determines whether or not the disease spreads through the body is the so-called *basic reproductive rate* R_0. This is defined to be the number of newly infected cells that result from one previously infected cell, considered during the initial period of infection, when there are very few infected cells. If this number is greater than 1, it means each infected cell will infect additional cells, and the virus will proliferate through the body. But if R_0 is less than 1, the single infected cell doesn't create additional cells that are infected, and the virus dies off."

"So what determines R_0?" I asked.

"Think about it this way. The rate at which one infected cell creates virions is κ. So we would expect κ virions to be created by this one cell in one time interval. But this one cell has an average life span of $\frac{1}{\alpha}$, so it should create $\frac{\kappa}{\alpha}$ virions in its lifetime. The rate at which those virions infect uninfected cells is $\beta b v$, where the v is the number of virions created by this one infected cell. But each virion has an average life span of $\frac{1}{\mu}$, so we get a total of $\frac{\beta b \kappa}{\alpha \mu}$ cell infections per infected cell. But remember, we are assuming this is at the beginning of the infection, so we know that the number of uninfected cells is $b = b_0 = \frac{\lambda}{\gamma}$ and therefore we get our conclusion that

$$R_0 = \frac{\beta \lambda \kappa}{\alpha \mu \gamma}.\text{''}$$

"Got it. So R_0 depends on all the constants. I guess you would expect that."

"Yup. And in the case of this virus, based on how the virus is spreading, the numerator $\beta \lambda \kappa$ is a lot bigger than the denominator $\alpha \mu \gamma$."

"So then what happens? Does the number of infected cells just continue to grow?" I asked.

"No, it can't. At some point, the number of uninfected cells shrinks to the point that there aren't many left to infect. So the system settles down to an equilibrium state."

"And what's that?"

"That's when the numbers b, i, and v stop changing. It occurs when all three of their derivatives are 0."

"A critical point."

"Exactly," said Jessie. "So if we set all three derivatives equal to 0, we can solve for the critical point (b^*, i^*, v^*). It turns out that at the critical point, the number of uninfected brain cells is

$$b^* = \frac{b_0}{R_0}.$$

(IX. Equilibrium value continued on p. 182)

"So what is probably happening is that R_0 is a lot larger than 1. The virus gets into the brain and does its damage, destroying a portion of the brain cells. But it doesn't destroy all of them. Eventually the system settles down to the equilibrium, leaving only the fraction of functioning brain cells given by $\frac{1}{R_0}b_0$. And the number of infected cells and virions also settle into their equilibrium values. This is what allows the zombies to survive. Their entire brain is never destroyed. A fixed portion always remains."

"So if R_0 was 3, the zombies would still have a third of their brain left."

"Right."

"That's interesting. So there's no reason to expect that the zombies are all going to keel over and die in the next couple of hours."

"No, I'm afraid not."

There was a pause.

"I think you're going to be okay," I said.

"We should know soon enough," said Jessie. "The minute I get weird, open the door and kick me out."

She stood up and motioned to the seat.

"Sit down," she said.

I shifted around her and sat down on the toilet seat cover. She sat down on my lap, and put her arms around my neck, leaning her head on my shoulder.

"I'm scared," said Jessie.

I wrapped my arms around her waist.

"I'm sorry I was so stupid," I said.

"About what?"

"About you."

She hugged me tighter. We waited....

"What's going on in there?" Gunderson's voice jerked us both awake. I looked at Jessie and she looked back at me. The small amount of light filtering in made it apparent it was dark outside.

"Not a zombie?" I asked.

106

"Don't think so," she replied. I leaned over and kissed her. She kissed me back.

"Are you okay?" she asked.

"I can't really feel my legs but, otherwise, I'm doing fine." She lifted herself off my lap, and I stood, as she shifted to the seat. My legs tingled. I looked at my watch and realized we had been in the port-a-potty for six hours.

"Hell," I said. "I have to call my daughter." I pulled out my cell phone and dialed. She didn't answer.

"This isn't good."

"Try someone else. Try Angus."

I punched his number but got no answer either.

"Maybe the cell phones aren't working anymore."

I tried my sister in California, but got no response there either.

"I have to get home," I said.

"Listen," said Jessie. "I don't hear the zombies." It was quiet outside the port-a-potty.

"Let's see what's going on," I said. I unlocked the door as quietly as I could and pushed it open a crack.

Strewn about on the ground were dozens of zombie bodies.

"Are they dead?" I asked.

"Unlikely," said Jessie. "More likely sleeping. That's another evolutionary trait that goes way back."

"So, you think it's okay to go out?"

"Yes, if we go very, very quietly."

I stepped out of the port-a-potty on rubbery legs, careful not to step on the zombies littering the ground. Jessie followed me out. I let the door shut very carefully so as not to make a sound.

"What are you two doing?" asked Gunderson, much too loudly.

"Shhh," said Jessie. "They're sleeping. You'll wake them up."

"Wouldn't want to wake the dead," said Gunderson, as he unsnapped the latch to his potty and swung open the door.

"Watch out," I said urgently, as Gunderson stepped out of the potty and planted the heel of his shoe right on the hand of one of the prone zombies.

Both Jessie and I sucked in our breath, waiting for the zombie to rise with a howl, and all the rest of the zombies to join in. But it didn't stir.

"Huh," said Gunderson as he nudged the zombie with his shoe. "Maybe they aren't sleeping at all. Maybe they're dead."

"Maybe the virus kills everyone after a few hours," I said hopefully.

Jessie leaned over the zombie, ready to leap back, but all was still. She had her ear on its chest when Gunderson gave it a good kick. She jerked back.

"Dammit, Oscar, was that really necessary?"

"Just checking," he said.

"It's alive," said Jessie. "I can feel it breathing. But it's an especially slow breathing."

"What does that mean?" I asked.

"I think it's hibernating."

"Hibernating?" said Gunderson. "Are you kidding me? It's spring. How do you know it's not sleeping?"

"In sleep, heart rate, breathing, body temperature are all essentially maintained. Relatively little noise and/or light is enough to wake you. But when animals hibernate, the heart rate slows, breathing slows, as the body needs less oxygen. The body temperature itself falls. It is much harder to wake an animal that is hibernating."

"So they're all suddenly hibernating?" asked Gunderson.

"I guess so," said Jessie. "Hibernation is another one of those traits that probably goes back millions of years. Humans don't do it, but a lot of other mammals do. Also lizards and other species that can't easily regulate their temperatures will behave similarly, only then it's called brumation."

"So what?" said Gunderson. "They'll wake in three months?"

"I don't know when they'll wake. Hibernation can last a day or six months. But I would suggest we take advantage of the opportunity while it exists."

"So now what?" said Gunderson.

"Now, I go home to take care of my kids," I said.

"What about you?" said Gunderson to Jessie. "We could find a safe place to hole up."

"I'm going with him."

"Oh," said Gunderson. "Things go well in the port-a-potty?"

Jessie ignored him.

"Well," said Gunderson, "I think I'll stick with you two for a while anyway. What's the plan?"

"Get down to Conrad Avenue, find an abandoned car with the keys in it, and drive up to my house," I said.

"Works for me," said Gunderson.

The three of us walked down through the parking lot toward the main road. None of the surrounding buildings had lights on. It looked like the power was out. I suspected it would be a very long time before it came on again.

There were bodies every few yards, some clearly zombies, from which we stayed as far as possible, and some clearly dead, with the stench of newly decomposing flesh beginning to rise off them.

When we reached Conrad Avenue, there was no shortage of available cars strewn over the road. But unfortunately their number meant that we couldn't possibly drive down the road.

"What now?" asked Jessie.

I looked up and down the road, and then spotted a building that might have a solution.

"Follow me," I said.

We crossed the road and trotted up half a block until we reached the Town Garage. We could see light streaming from the windows.

"They must have a generator that kicks in automatically when the power goes out. They need this building to remain functional during storms."

The side door was unlocked, and we slipped inside. Stepping through the main office, we entered the garage area, which was well lit.

"I get it," said Gunderson, as he smiled up at a huge snowplow. There was a row of five of them. I walked down until I stood in front of the one that had a big V-plow on the front. The tires were a good four feet in diameter.

"This one," I said.

Grabbing hold of the door handle, I pressed the button, hoping it wasn't locked. The door swung open. Hoisting myself up, I slid into the driver's seat.

"Too bad," I said. "No keys."

"I'll check in the office," said Jessie.

I waited in the truck, as Gunderson wandered around the garage.

Suddenly Jessie screamed.

chapter eight

HOUR 18

"Dammit," I said, feeling stupid for not being more careful. I leaped out of the truck and pulled the gun out of my pocket, as I sprinted toward the office. Gunderson was right behind me, a stricken look on his face.

Jessie was up against the wall, with her arm out straight and her fingers around the neck of a large guy who must have worked at the garage earlier in the day. He was doing his best to get his mandibles around her face.

I came up beside him, put the gun to his temple and pulled the trigger. The bang echoed around the building. Blood sprayed out of the far side of his head and he collapsed to the floor.

"You okay?" I said to Jessie, looking her over for bites and scratches.

"I think so," she replied. She gave me a hug. "Thank you."

Gunderson stood in the doorway. "Well, I guess they aren't all hibernating," he said.

"Yeah," I said. "It's funny he was awake."

"Maybe not so funny," said Jessie. "Notice it's warmer in here than outside?"

"Yeah," I said. "I assume the heat's on in here because of the generator."

"And when evening came on," continued Jessie, "the temperature dropped outside."

"Yeah?"

"Yeah. I think the zombies can't regulate their temperatures. So with a drop in temperature, they do what a lot of animals and especially cold-blooded creatures do, they hibernate."

"So our friend here wasn't hibernating because it's not cold in here?" said Gunderson.

"Exactly," said Jessie.

"But is he the only one in the building?"

As if to answer the question, we heard a low moan from the garage.

"Shut the door to the garage," I said to Gunderson. As he did so, three zombies rounded the front of the nearest snowplow. We watched through the window in the door as five more joined them. They took turns slobbering on the glass pane.

"Well, so much for riding one of the snowplows out of here," said Gunderson.

"Can't you shoot them?" suggested Jessie.

"Out of bullets. That was the last one," I replied. "We're stuck."

"Maybe not," said Jessie.

"What do you have in mind?"

"First, we need the keys."

"Here," I said, pointing to a board with sets of keys hanging on it. "But which set of keys is it?"

"Assuming the trucks are in order," said Gunderson, "the truck we were looking at was number three, which should be this set of keys here." He plucked the keys from a hook.

"Good," said Jessie. "Now, all we have to do is lower the temperature in here."

"How do we do that?" asked Gunderson.

"First we open the garage doors," I said. I hit three buttons next to the desk and we heard the garage door mechanisms lift the three doors. It would have been nice if the zombies in the garage had then wandered off, but they could see us, and they weren't going anywhere.

"Now what?" said Gunderson.

"Now we kill the heat," I said, as I walked over to the thermostat on the wall and turned the heat all the way down.

"Okay," said Gunderson, "so now how long do we wait?"

"Well," I said, "we know how the temperature of an object changes with time."

"You mean Newton's Law of Cooling?" said Gunderson.

"Yes," I said. "We can assume that the temperature in the garage will quickly level off at the outside ambient temperature, which, according to the thermometer in that window, is about 50°. Figure the zombies were in here long enough that their temperature is the temperature that the building was, which according to the thermostat was 70°."

"Not 98.6°?" said Jessie.

"Not if we're right about them being unable to regulate their temperatures. Hypothetically, we evolved from cold-blooded animals. So it depends on how far back the virus sends us evolutionarily. But given the hibernation behavior, it seems plausible that the zombies can't generate their own heat."

"Okay," said Jessie.

"So then if they don't produce their own heat, their temperature will drop according to the differential equation that dictates temperature change of an object placed in a medium of a different temperature."

113

"You mean like a zombie of temperature 70° placed in a room of temperature 50°."

"Exactly. Newton's Law of Cooling says that the rate of change of the temperature of a body is proportional to the difference between the current temperature of that body and the ambient temperature."

"Can you repeat that in math?" asked Gunderson.

"It means that if T is the zombie's temperature, given as a function of time t,

$$\frac{dT}{dt} = -k(T - T_G)$$

where T_G is the temperature of the garage." I wrote it in the corner of an invoice sitting on the desk.

"So we take T_G equal to 50°," said Jessie.

"Right."

"But why is that the differential equation that gives the zombie's temperature?" asked Jessie.

"It makes sense. You would expect that the bigger the difference in temperature between the body and the surrounding temperature, the faster the temperature of the body would drop."

"True," said Jessie. "But can we find the function that satisfies that differential equation?"

"Yup. We can use a substitution to figure it out. Just let $y = T - T_G$."

"Why would you do that? Adding a new variable just makes it more complicated," said Gunderson.

"Give me a second," I said. "Notice that $\frac{dy}{dt} = \frac{dT}{dt}$. So substituting into the original equation by replacing the expression $\frac{dT}{dt}$ by $\frac{dy}{dt}$, and replacing $T - T_G$ by y, we get $\frac{dy}{dt} = -ky$."

"Isn't that the differential equation that gave us exponential growth?" asked Jessie.

"Pretty close," I said. "Except because of the extra minus sign, it gives us exponential decay."

"Meaning?"

"Meaning we get as a solution $y = Be^{-kt}$, where B is a constant."

"But we don't care about y," said Gunderson.

"Right, so now we substitute back in for T and we get $T - T_G = Be^{-kt}$. So

$$T = T_G + Be^{-kt}.\text{"}$$

"That was an unnecessarily complicated way to solve it," said Gunderson. "You could have just taken the original equation and separated the variables." He wrote the following on the invoice:

$$\frac{dT}{dt} = -k(T - T_G)$$

$$\frac{dT}{T - T_G} = -kdt$$

$$\int \frac{dT}{T - T_G} = \int -kdt$$

$$\ln(T - T_G) = -kt + C$$

$$T - T_G = e^{-kt+C}$$

$$T = T_G + e^{-kt+C}$$

$$T = T_G + Be^{-kt} \quad \text{where} \quad B = e^C.$$

"Same answer," he said.

"I was doing it in a way that related it to what we had previously talked about. I thought that would make it clearer."

"Why would that make it clearer? Now she'll confuse the two different situations."

"Guys," said Jessie. "Do we really need a pedagogical debate right now?"

"It's not me," I said. "It's Gunderson."

Jessie sighed. "Can we get back to Newton's Law of Cooling? What are the constants k and B?"

"Well, B is easy," I said. "At time $t = 0$, let's say that the zombie's temperature is T_I, where the I stands for its initial temperature. When $t = 0$, we get

$$T_I = T_G + Be^{-k0} = T_G + B.$$

So B must equal $T_I - T_G$. Therefore we have

$$T = T_G + (T_I - T_G)e^{-kt}.$$

So that's our equation. In our case, $T_G = 50°$ and $T_I = 70°$, so we have

$$T = 50 + 20e^{-kt}.$$

"It looks like this," I said as I drew Figure 8.1 on the invoice. "Which particular curve it is depends on what k is.

"Notice that in all cases, when $t = 0$, the temperature of the zombie is the initial temperature, 70°. And as t approaches infinity, the temperature of the zombie approaches 50°."

"I see," said Jessie. "As t gets large, $e^{-kt} = \frac{1}{e^{kt}}$ approaches 0."

"That's all great, but what's k?" asked Gunderson.

"That depends on the object that is cooling."

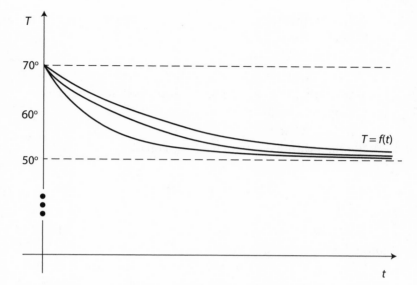

Figure 8.1: Zombie with initial temperature 70° cooling in a room of temperature 50°. Each curve corresponds to a different value of k.

"So you're telling me you went through all this only to tell me you don't know k," said Gunderson.

"Well, usually in determining the time of death of a body, for example, the coroner takes a couple of readings of the body temperature, a set time interval apart, and that additional information determines k."

"I see," said Gunderson. "So how does the coroner take those temperature readings?"

I looked down at the floor.

"Usually rectally."

Gunderson laughed. "So you're suggesting we go out there in the garage and rectally measure the temperature of one of those zombies who's slobbering on the window? Great plan, Williams. I'm sure they'll be happy to go along."

"Well, ummm. . ."

"I know k," said Jessie.

"How's that?" I asked.

"I teach a course on forensic biology. One of the projects has to do with how long it takes a dead body to cool. For a male of average build, you can take $k = .2$."

I smiled. "Thanks."

"So then with $k = .2$, we have the equation

$$T = 50 + 20e^{-.2t}.\text{"}$$

"Yes, but we don't know what temperature puts them into hibernation," said Gunderson.

"They went into hibernation when it was warmer than this, so let's take a guess and say 60 degrees."

"Then how long will it take for them to drop to 60 degrees?" asked Jessie.

"We set $T = 60$ and solve for the time t."

I wrote on the invoice:

$$60 = 50 + 20e^{-.2t}$$

$$10 = 20e^{-.2t}$$

$$\frac{1}{2} = e^{-.2t}$$

$$\ln\left(2^{-1}\right) = -.2t$$

$$t = -5\ln\left(2^{-1}\right) = -5(-1)\ln 2 = 5\ln 2.$$

"Okay, so what is that?" asked Gunderson. "We don't have a calculator."

"You just need to know $\ln 2$, which is 0.693," I said.

"How could you possibly know that?" asked Jessie.

I shrugged. "It comes up enough that I know it. So I'm a nerd. Then $t = 5 \ln 2 = 5(0.693) \approx 3.5$. So about three and a half hours."

"All right," said Gunderson. "We can wait that long."

"Maybe you can," I said. "But I can't. I have to get to my kids."

"What can we do?" asked Jessie.

"Get ready to close the garage doors." I opened the door to the outside of the building and slipped out. Keeping close to the building, I crept around to the open garage doors. Stepping out in front of the opening, I yelled, "Hey, you crazies, come and get me."

The zombies turned from the office door and proceeded to lurch after me. Once they were all out of the garage, I yelled to Jessie, "Shut the doors."

I circled back to the door I had exited with a parade of zombies shuffling behind me. Reentering the building, I slammed the door in their faces. Now we had eight zombies pounding on a different door.

"Okay," I said. "Let's get out of here."

We returned to the garage, and I climbed back up into the driver's seat of the truck.

"Uh-uh," said Gunderson, dangling the keys. "I'm driving."

"The hell you are," I said. "I'm driving. I need to get to my kids."

"Oscar, this was Craig's idea," said Jessie.

Gunderson hesitated, and then said, "Oh, all right," as he tossed me the keys.

I started up the truck. "Gunderson, you need to go back to the office and hit the garage door opener and then hightail it back here."

"Not on your life," said Gunderson. "You'll take off without me."

"You are such an ass," I said.

"I'll do it," said Jessie.

As she returned to the office, Gunderson came around the far side of the cab and hopped in.

"Gunderson," I said. "Do you really think I would have left you?"

"You have every reason to do so," said Gunderson. "You see me as a challenge to your relationship with Jessie."

"Are you kidding?" I responded. "Do you really think I'm worried about that now?"

"Let's be realistic," said Gunderson. "Who knows how many humans are going to survive this? It's not like we'll be able to go on match.com and find a mate. The pickings will be very slim." The garage door started to rise.

"You are a complete idiot," I said to him, as Jessie ran across the front of the truck and hopped in next to Gunderson. I put the truck into gear.

One zombie who had heard the noise of the door rising gave a rusty moan and lurched in front of the snowplow, with arms extended. He was missing a good chunk of the left side of his face.

"Sorry, fella," I said as I stomped the accelerator to the floor and let the clutch go.

The truck jumped out of the garage, and the zombie hit the plow with a thunk. He disappeared beneath the truck, as I pulled out onto Conrad Avenue. I shifted into second gear. Up ahead were two cars that had collided and that together blocked both lanes. Trees to either side prevented us from circumventing the cars.

"Here we go," I said, as I shifted into third, and stomped the accelerator to the floor, aiming the point of the V-plow for the back end of the right car. We hit it with a loud crunch, as sparks flew up and the crumpled car slid across the plow and to the right. The truck barely slowed down.

"All right," I said.

"We're in business," said Gunderson.

As we drove down Conrad Avenue, we could see the results of the day's mayhem. Some houses remained intact, but many others had broken glass or were in smoldering ruins. Many bodies littered the ground, but it was impossible to tell if they were hibernating zombies or dead humans. We could swerve around many of the cars blocking the road, but every hundred yards or so, we had to ram our way through. Several times, we had to stop and climb out of the plow to survey the situation and determine the best way to ram a path through. Eventually, we reached the end of Conrad Avenue and turned onto North Hammond Street. It was eight miles down this road before we would reach Cobble Road, and another two miles up it to my house.

There were fewer cars on North Hammond, but it was still slow going. After successfully navigating the first three miles, we came to the railroad crossing and what was clearly the end of the line. A train blocked the way forward. It had derailed, and many of the cars lay on their sides. There was no possibility of ramming our way through this.

I pulled the truck up to the side of the road.

"What do we do?" asked Jessie.

"From here, we hike," I said.

"How far?" asked Gunderson.

"By the road, it's about seven miles, but through the woods, I would guess around five miles."

The sky was just beginning to lighten.

"The zombies will be waking soon," said Gunderson.

"Yup," I said. "So the sooner we get moving the better." I opened my door and hopped out of the truck. Jessie did also. But when Gunderson jumped down, he let out a yelp and fell to the ground.

"What happened?" I asked as I came around the truck.

"He twisted his ankle," said Jessie as she knelt beside him.

"God, it really hurts," said Gunderson as he held his left ankle. "I hope I didn't break it."

"Now what?" said Jessie. "He won't be able to hike five miles."

"I have to get to my kids. You stay with Gunderson in the truck. You'll be safe there for the time being. I'll go get my kids and come back here."

"No," said Jessie.

"We can't just leave him," I said.

"Yeah," said Gunderson. "I can't even walk."

Jessie frowned.

"Okay," she said. "We'll wait for you here."

"What if you don't make it back?" asked Gunderson. "We would never know and we'd just be stuck waiting here."

"Give me four hours. If I'm not back by then, take the truck back to campus."

"Give me the keys," said Gunderson as Jessie helped him to his feet.

I ignored his proffered palm and handed them to Jessie.

"I'm sure your kids'll be okay," she said. "See you in a bit."

"Yeah, see you later, Williams," said Gunderson. He was supporting himself with an arm around Jessie's shoulder.

I hesitated for a moment, thinking over what I could say. But none of the options that came to mind seemed appropriate so I just shook my head as I turned and walked away. Ducking under the coupling between two of the railcars,

I headed down the road another half mile, until I found the trail that lead up into the woods and that would take me to my house. An hour and a half later, I could see the mist rising from the roof of my house as the sun burned it off.

From the outside, the house looked unharmed. No broken glass, no smoke. I walked up to the front door and took a deep breath. Trying the handle, I was dismayed to find it unlocked. I pushed the door open and entered cautiously. Ellie was standing in the living room, facing away from me.

"Ellie? Ellie, are you okay?"

She turned, a dreadful look on her face.

"Honey, please tell me you're okay."

She charged at me. I stood frozen, unable to protect myself, unable to move as she leapt upon me. I waited for the inevitable bite, but instead all I heard was, "Dad! Dad! You're alive." She was holding onto me with all her might and crying uncontrollably. I wrapped my arms around her.

"Oh, honey," I said. "You're okay."

"Yeah, Dad, but bad things happened."

"What kind of bad things?"

"Connor wouldn't stay here. He said he was going up to Jeff's house."

"What? Why didn't you stop him?"

"I tried, Dad. I told him you said to stay, but he ignored me. He said I wasn't the boss of him."

"Damn it," I said.

"But then, when he went out the front door, Mr. Winsted came out of the woods. There was something really wrong with him. He started to chase Connor. And Connor started screaming. I didn't know what to do. That's when Rusty ran out of the house. And Rusty could tell Mr. Winsted wanted to hurt Connor so Rusty attacked him."

"Oh, God."

"Yeah, Rusty got him by the leg and was growling and tearing at his leg. Mr. Winsted fell down, and they started wrestling around. Then Mr. Winsted bit Rusty on his thigh!"

"He did? Where was Connor during this?"

"He ran up the road to Jeff's house."

"So then what happened?"

"Rusty got Mr. Winsted by the throat, and then they rolled down the hill at the edge of the yard. I was afraid to go look, but then Rusty came back up the hill. He was pretty scratched up. I don't know if Mr. Winsted is still down there."

"Where's Rusty?"

"He's in the study. I moved his dog bed and water dish in there, and then I shut the door."

"Smart, honey. We don't know if dogs can get the virus. Have you been in there?"

"No."

"Okay. I'll see how he's doing. Wait here."

I stood outside the door to the study and listened, but it was quiet.

"Rusty, you doing okay?" I said through the door.

I heard a whimper. Cracking the door open, I peered in. Rusty was sitting on the other side of the door, with his head cocked to the side, looking at me quizzically.

I pushed the door open. "Hey, Rusty," I said. He came over and nuzzled against my leg.

"You did a good job, Rusty. You're a good dog," I said as I patted him on the head. I pointed at his bed. "Lay back down now." Rusty curled back up on his bed, and I shut the door.

"We'll check on him in a bit, but I think he's going to be okay," I said to Ellie. "But now I've got to find Connor. I want you to stay here."

124

"No, Dad. I'm not staying. I want to go with you."

"It's much safer here."

"I don't care. I'm going with you. He's my brother."

"Honey, I can't protect you at the same time as I'm trying to rescue him."

"Dad, I've been able to run faster than you for the last three years."

That was true. She was much more of an athlete than I ever was. I went into the mudroom and grabbed two of Connor's bats.

"Here," I said. "Anybody comes close, swing for their head."

"Are you serious?" asked Ellie.

"Dead serious," I replied.

We walked outside and I scanned the surrounding woods for any sign of zombies, but it looked clear.

"Keep your eyes open for trouble," I said as we walked down to the end of the driveway and up toward Jeff's house. I didn't know if the zombies were warmed up yet or not.

I could see the damage to my neighbors' houses. Several bodies lay on the lawns. I didn't want to know who it was. Jeff's house was the last on the street, built into the hillside with the highest elevations, affording an excellent view down on the valley. We climbed the driveway. The house looked undamaged from the outside.

"Ellie, I want you to wait here. I'll go in the house and see what I find."

I left her on the driveway and walked up to the front door. It seemed silly but I knocked. There was no response. After waiting twenty seconds, I pushed down on the door handle and the door swung open. I had the bat at the ready as I stepped inside.

"Connor?" I called. "Are you here?"

125

I was met with silence.

"Doug? Karen? Are you here? Jeff? Anybody home?" I yelled.

I walked through the foyer to the living room. It was at this point that I spotted the pool of blood in the center of their white rug. I couldn't help remembering all the times I had taken off my shoes before I came in this house, because of that white rug. The most disturbing aspect was that there was no body to go with the blood.

"Connor?" I called again, more softly.

Suddenly I heard Ellie scream. I could see her through the window and her back was to a tree. Lurching toward her was Jeff.

"Hit him," I screamed as I bolted for the door. "Hit him, Ellie!"

"It's Jeff," she called back. "I can't hit him."

"You have to, honey!" I screamed.

I had reached the front door and could see that I wouldn't get there in time.

"Remember softball?" I yelled. "Swing for the bleachers."

Jeff was almost upon her when I saw her wind up, and after giving me a terrible look, she swung with all her might.

I heard the crack as her bat connected with Jeff's head. Jeff went down hard. I reached her and wrapped my arms around her.

"Good job, honey."

"I think I killed him," she said, starting to sob.

"It's okay, honey. It's okay."

"He came around from the back of the house. I didn't have time to run."

"You did the right thing, honey," I said. "Let's take a look back there." I was deathly afraid of what we might find.

We crept along the side of the house until we reached the back corner and then we peeked around it into the backyard.

Doug and Karen were at the base of the tree that held Jeff's tree fort. Doug was trying desperately to pull himself up the set of short 2 × 4s that were nailed into the tree to make a ladder. But each time he would get up a step or two, his foot would slip and he would tumble down to the ground. It reminded me of the coordination problems the zombies suffered from in the auditorium. A head appeared in the window of the tree fort.

"Connor, is that you?" I yelled.

Karen and Doug turned from the tree fort at the sound of my voice.

"Dad, help me. I'm up here. I've been stuck up here for hours. Mr. and Mrs. Varden are down below and they want to eat me."

"Connor, it's okay," I said.

"I told you not to leave the house, Connor," said Ellie.

Karen and Doug started staggering in our direction, realizing that finally here was a meal they could get to.

"Going to need your help, honey," I said.

The Vardens advanced on us, drool rolling down their chins. Doug was in the lead, so I stepped up and cocked the bat. But just as his head came into range and I swung, he raised his arm to reach for me. I could hear the bones on his forearm crack, but the bat missed his head entirely. The momentum of the blow made him stumble, but he didn't lose his balance. He turned to attack me and because of my follow-through, I couldn't get my bat back around in time.

It was at this moment that Ellie let her bat fly on the back of his head, and he went down.

As she was looking at her handiwork, I yelled, "Duck!" She understood my urgency and dropped to the ground as I swung, catching Karen dead on her chin. Her neck snapped back, and she keeled over backward.

"Good job, honey," I said, patting her arm. "I'll go through the apocalypse with you anytime." I'm a firm believer in positive reinforcement. We walked over to the tree fort.

"Connor, it's clear," I said. "Come down."

Connor opened the trapdoor and climbed down the 2 × 4 ladder.

"Thank God you're okay," I said, as I gave him a huge hug.

"I was smart, Dad. I went up in the tree fort. They can't climb worth a damn."

"Watch the language," I said without thinking.

"Come on, Dad. Ellie just cracked open Mr. Varden's skull and you're giving me trouble about swears?"

"Okay, but I know Ellie told you to stay home. You should have listened to her."

"Sorry, Dad," said Connor.

I patted him on the arm. "Forget it. It's not like I'm going to take away electronics. Come on, we gotta go."

We trotted down the driveway and back along the road to our house. One or two neighbors drifting through the woods spotted us and began their slow pursuit, but we made it back to our house without incident. I locked the door behind us.

"Okay, guys, here's the deal. I'm going to fill three backpacks with food and drink. Go upstairs and put on a bunch of layers. Some long underwear, two pairs of jeans, a sweater or two. Then come down and put snowpants and a jacket over that."

"Dad, we'll be way too hot," said Connor.

"Better hot than dead," I replied.

I grabbed the backpacks hanging on the hooks in the mudroom and filled them with crackers, chips, water bottles, Gatorade, cheese, snacks, some leftover pizza, and anything else that would help to keep us going. I threw in some flashlights, batteries, matches, and masking tape.

Then I went upstairs and put on two pairs of jeans and two sweaters over my shirt. I came down to the mudroom and added a denim jacket over the top.

When the kids came down, I sent Connor to get his football helmet and I put a bicycle helmet on Ellie. Then I had them put on gloves and I took the duct tape and taped the ends of the sleeves of their jackets to their gloves. They looked like the padded assailants used to train police dogs.

"I'm already hot," said Connor. I pulled the zipper of his jacket down.

"Once we get going, we'll have to zip that."

I heard a bark and went to the study door. Not hearing anything unusual, I inched the door open and peaked inside. Rusty was sitting there, an expectant look on his face. He thought he was overdue for a walk.

"Okay, Rusty. You can come." He wagged his tail as he followed me out.

"Are you bringing his food?" asked Ellie.

"He'll be just as happy with crackers," I replied. I handed each child a backpack and slung my own over my shoulders.

"We have to hike about five miles. Keep moving at all times. Remember, we're faster than they are."

I handed Ellie and Connor the bats and pulled a driver out of my golf bag.

"If you have to fight a zombie, aim for the head and hit them as hard as you can. Otherwise they'll just keep coming.

But if you can get away by running, do that. We're all a lot faster than they are. Okay, let's go." I opened the front door a crack and looked out.

Two zombies had followed us home, one of whom I didn't recognize and the other of whom was Mrs. Wimmer, our eighty-five-year-old neighbor. Both were facing the other direction and were several yards from the door.

"Ready?" I said. "When I say go, I want each of you to sprint down the driveway, and then down the road. I'll be right behind you."

The two of them nodded and I softly said, "Go." With Ellie in the lead they ran down the driveway. The zombies realized they were there and started after them. "Come on, Rusty," I said. I came up behind the first zombie and swung the bat into the back of its head, knocking it flat on its face. I couldn't bring myself to swing at Mrs. Wimmer, so I just waved the bat at her as I ran by.

Rusty and I overtook the kids twenty yards down the road. We trotted along until I pointed us into the woods. As we jogged down the old Greylock trail in all the clothing, I started sweating almost immediately.

"Dad, I'm too hot," said Connor, panting.

"Complaining already," said Ellie.

"It's hot," said Connor. "Do we have to run?"

"No. For the time being, we can walk, but make it a fast walk." I was worried that we wouldn't make it back to the truck before Jessie and Gunderson headed back to campus.

Rusty trotted down the trail thirty yards ahead of us. I looked back and spotted several zombies who had picked up on us and were making a slow pursuit.

"Keep it moving," I said. The trail took us deeper into the woods. If we did run into trouble, there would be no doors to duck through. Suddenly, up around the bend, Rusty started howling.

Ellie stopped. "What do we do, Dad?"

I pointed behind us at the zombies following us down the trail.

"We keep going," I said. "But get ready to run. I'll take the lead."

We came around the bend to find Rusty barking at three zombies. They were doing their best to ignore him as they scrabbled at a tree. On a branch above them crouched a young boy who was at most ten years old.

"I know him," said Connor. "His name's Cole. He lives on North Main."

"We can't leave him," said Ellie.

"I know," I said. "I have number one, the big guy on the left. Ellie, you take number two, the woman in the middle, and Connor, you take number three, the woman on the right."

"You mean I really have permission to hit her?" he asked incredulously.

"Yes, you do. And I would really like it if you hit her really hard."

"You got it, Dad."

"No noise until we are right on them," I said. We crept up the trail until we were only a couple yards away, when one of them noticed us and turned.

"Now!" I yelled. We sprang forward. I swung hard, careful to avoid the zombie's arms, and I caught it with a sharp crack right in the temple. It collapsed to the ground. Connor connected with his on its forehead, and it sank to its knees, but didn't go all the way down. Ellie's zombie was enough behind Connor's that Ellie couldn't get a good swing in. Her zombie pushed past Connor's, which was just

131

struggling back to its feet, and fell upon Connor. The two tumbled to the ground, the zombie trying its best to bite through Connor's football helmet, slobbering all over its shiny surface in the process.

"Get off me, you freak!" yelled Connor.

Connor's zombie managed to regain its feet and it fell upon the two of them, doing its best to chew through the sleeve of Connor's coat. Snarling, Rusty leaped in and started pulling at its leg. I stepped forward and swung, catching Connor's zombie on the back of its head, and it collapsed to the side.

Ellie took the next swing and the third zombie fell beside it.

"Connor, you okay?" I asked.

"Yeah, Dad. The gear worked great. They couldn't get me." He was grinning.

"They wouldn't have gotten you in the first place if you'd hit the first one right," said Ellie.

"It's not like you hit yours," said Connor as he stood up and picked up his bat. "Mine must've had an extra thick skull. Kind of like your skull."

"Can I come down?" asked the kid.

We all looked up.

"Oh, yeah," I said. "Cole, right?"

"Yeah," he said, as he climbed down the branches to the ground.

"Cole, where are your parents?" I asked.

"That's them right there," he said, pointing at two of the zombies we had just clubbed to the ground.

"Oh, ummm, I'm really sorry about that," I said, shifting the bloody bat behind my leg.

"They were trying to kill me. I already saw them eat some other people. They're better off dead."

"Okay," I said. "I guess we'll deal with the psychological trauma later. Why don't you stick with us, at least for the time being?"

Cole nodded, and the four of us set off down the trail, with Rusty again taking the lead. In another forty minutes, we came out on North Hammond Road.

"Wow, a train wreck," said Connor as we approached the railroad crossing. I could see the truck on the far side of the train, but hesitated to call out for fear of attracting any zombies in the vicinity.

We ducked under the coupling and I stepped up on the truck runner and peered inside the cab. Jessie was asleep with her head against the window. Gunderson was splayed out across the seat with his head in Jessie's lap.

"Seriously?" I said out loud. I rapped on the window, and both were startled awake. Gunderson sat up, as Jessie unlocked the door.

"Oh, thank God," she said. "We've been so worried. You're okay. And Ellie and Connor. It's so good to see you!"

"What are they dressed as?" said Gunderson. "Sumo wrestlers?"

"Jessie, this is Cole. He's joining us." Cole nodded in her direction.

I pointed to the metal rungs that led up the side of the open-box truck bed.

"All kids climb in the back."

"Really?" said Connor. "We get to ride in the back?"

"Really," I said. "Come on, Rusty. You ride in front with us."

As the kids clambered into the box, I let Rusty climb in and settle at Jessie's feet. I pulled myself up into the driver's seat and Jessie handed me the keys. Rolling down the window, I yelled back, "Everybody, sit down, and stay down."

133

Then I started the truck and headed back toward the college. The path we had cleared earlier was still open so it didn't take us long to get there. We passed a few wandering zombies who did their best to get our attention but they soon disappeared behind us. I pulled the truck up to the curb in front of the Science Center and turned the engine off. It's not like Security would give me a ticket.

"Gunderson, you wait in the cab," I said. "With your ankle, it would be too dangerous for you to come in with us."

"Oh, no," said Gunderson. "You're not leaving me here. I'll keep up."

"Oscar, we won't leave you," said Jessie. "We're just going to get the others. It's safest for you to stay in the truck."

"Okay, " said Gunderson."But make it quick."

"Stay," I said to Rusty, and he curled up under Gunderson's feet.

There weren't any zombies in sight, so the rest of us clambered out of the truck and made our way toward the doors of the Science Center. I noticed the lights were on, and remembered that the Science Center had its own generator to protect scientific equipment and ongoing experiments from power failures. That meant the heat had been on. Whatever zombies had been in the building, they had not been hibernating.

We heard a banging noise and looked up. There was a student at a second floor window motioning to us. It was Thora. She pulled open the bottom section of the window and yelled out desperately.

"Help me!" she howled. "I'm trapped in here!"

"Hang on. We're coming," I yelled back.

"Be careful," I said to the others, as I pulled open the door to the Science Center and we entered.

The stairwell was deserted, so we ascended cautiously until we reached the second floor. I peered out the window in the fire door, but couldn't see the door to the office that contained Thora.

"Okay, I said. "Connor and Ellie, you have the best protection. You two come with me. Jessie and Cole, wait here."

I held my golf club at the ready and pulled open the fire door. The three of us slipped through, and then keeping to the wall, we quietly moved down the hall. Ahead of us, I could hear movement. As we continued forward, we heard the characteristic moan of a frustrated zombie.

We turned the corner, and there, scrabbling at the door to the office containing Thora, was Megan. Her head was at a very unhealthy angle relative to her shoulders. Her clothes were torn and substantially more disheveled than the last time I had seen her. She had lost her shoes and the pearl necklace she had always worn.

She turned as we approached, and at that instant, Thora threw open the door to the office.

"Thank God you're here," she said. "This bitch just won't go away."

"It might have been better to wait a minute to open that door," I said. Megan looked first at us, and then at Thora, as if considering her options. I wondered for just a moment if there was something left of Megan inside this creature, if the virus didn't actually destroy all those parts of the brain that had made up the individual who had been Megan. And I wondered if somewhere deep inside she recognized me and she recognized Thora and that might influence her decision as to which way to go. At that instant, she turned from us and lunged at Thora. Thora shrieked and dodged back into the office, slipping behind the desk.

"She always hated me!" she howled. Megan tried to grab her across the desk, but Thora leaned far enough away to stay out of reach. Megan didn't seem to have enough sense to go around the desk, and just kept reaching. Finally she put a knee up to climb over the desk.

"Megan," I said loudly enough to make her turn. "Time to call it a day."

She stopped trying to climb over the desk and turned to face me as I stepped forward. Her face broke into a hopeful grin, and she held out her hand as if proffering it for a greeting kiss. Then with a snarl she lunged forward and I swung the club, catching her solidly on the cheekbone. Her already weakened neck snapped, and she crumpled to the floor.

"Have you been working on your stroke, Dad?" said Connor.

"Not appropriate," I replied, a wave of sadness rolling over me.

Thora came around the desk and wrapped her arms around me.

"Thank you, Professor Williams. Thank you."

"No problem," I said awkwardly. Both of my kids rolled their eyes.

"Come on, Thora. You'd better stick with us."

chapter nine

HOUR 24

I banged on the door.

"Hey, it's us," I said. "Open up."

Angus cracked the door and then threw it open.

"Thank God you're here," he said.

"Are you okay?" I said, as we herded in. "Did the zombies get in? Did someone get bitten?"

"No," replied Angus, "but do you have any idea how excruciating it is to be the only student trapped in a room with a woman whose only mission is to counsel you?"

Sylvia gave him a resentful look. Marsha hugged Jessie and me in turn.

"So good to see you all," she said.

Ortiz smiled at us.

"How you doing, Raphael?" I asked.

"I'm better," he replied. "Just needed some rest."

"Found the candy machine, I see," I said, looking over the chip bags and candy wrappers strewn over the desk. "And Marsha, how'd you get the sneakers?"

She patted Angus affectionately on the shoulder. "Angus got them for me out of my office. If you ever have him in class again, you better give him a good grade, or you'll have me to answer to."

If anybody has anyone in class again, I thought to myself.

"I have something else to show you," said Angus to Jessie and me. "Come with me." He cracked the door open and then motioned for us to follow him. The three of us walked down the hall to the big freezer where the low temperature chemicals are kept.

"I've got some zombies in here," he said, as he pulled on the handle.

"Angus, no," I said, as I cocked my golf club.

The door swung open. Lying on the floor were three zombies, one of whom was a senior I recognized from my topology class. The other two were members of the Physics Department who worked together on low temperature physics.

"Are they dead?" I asked.

"Nope," said Angus. "As far as I can tell, they're sleeping. It's like the cold put them into hibernation."

"Yeah," I said. "We came to the same conclusion. They hibernate when it gets cold."

"So I figure we just have to lay low during the day, and then we can come out at night, when they're hibernating," said Angus.

"We're only safe on the nights the temperature drops to below 60°," said Jessie. "Really, we need it to drop lower, so there is a substantial amount of time the zombies are hibernating. Say down to 50°."

"How often is that this time of year?" asked Angus.

"I'm a weather geek," said Jessie. "So I can tell you. The average low temperature per twenty-four-hour period in Westbridge in April is 56°. And the standard deviation is 5.6°. You remember the 68-95-99.7 Rule, Angus?"

"Yeah. It says that 68.3% of the time, the temperature is within one standard deviation of the mean. So that means that 68.3% of the time, the low temperature is between $56 - 5.6°$ and $56 + 5.6°$. So it's between 50.4° and 61.6°."

"Yes," said Jessie. "So the remainder of the time, which is 31.7% of the time, the temperature is either greater or less than that."

"So half of that 31.7%, which is about 16% of the time, the low temperature is around 50° or less."

"Exactly."

"That's not much of the time," said Angus.

"No, it isn't," said Jessie. "Only one out of every six nights. And as summer comes on, it will only get warmer."

We walked back to the office and secured the door.

"So what's the plan?" asked Marsha.

"We need a safe place for the long term," I replied.

"I thought the logistic model said there were no safe places. Eventually the zombies win out and there are only zombies left," said Angus.

"It's not that simple," said Jessie.

"But you said it," said Marsha.

"I also said it was a simplified model."

"Well, we've got time," said Angus. "What's the complicated model?"

"We could use the Lotka-Volterra model," said Jessie.

"The what?" asked Angus.

"It's what's called a predator-prey model, where you have one species that's the prey and another species that's the predator."

"Certainly seems appropriate," said Ortiz. "How does it work?"

"Well, we let H be the number of uninfected humans and we let Z be the number of zombies. Let's think long term. Initially, the number of humans will drop precipitously."

"Maybe already has," said Ortiz.

"Yes," continued Jessie. "So then the remaining humans will have plenty of food."

"How do you figure that?" asked Marsha. "It's not like farmers will be planting and sowing any time in the near future."

"Yes, but think about all the canned goods that exist in stores and warehouses right now. With the diminished population, we should be fine for quite a while."

"Yeah, and plenty of candy machines out there," added Angus.

"Right," said Jessie. "So eventually, the remaining humans have the potential to start having babies again, and with enough resources, and temporarily ignoring the presence of the zombies, we would expect to see exponential growth in the population."

"You mean like the first differential equation we talked about for zombies," said Angus. "The one that says $\frac{dZ}{dt} = kZ$."

"Yes, but now it would be $\frac{dH}{dt} = \alpha H$, where α is the growth rate."

"But then what happens when we include the zombies?" asked Marsha.

"Then, the zombies will kill off some humans and also zombify some others at a rate proportional to how often the two groups meet. The more zombies there are and the more humans there are, the more often the two groups meet. So we get a term that corresponds to the loss of humans from those meetings of the form $-\beta H Z$, for some fixed constant β.

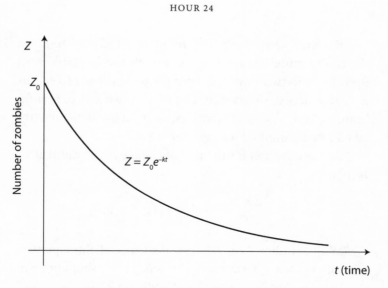

Figure 9.1: The decay in the number of zombies if there are no prey.

So our differential equation becomes

$$\frac{dH}{dt} = \alpha H - \beta H Z.\text{"}$$

"And what about the zombies?" I asked.

"First, think about what would happen to them if there were no humans. The number of zombies would exponentially drop off. They would begin starving to death. So the equation in that case would be

$$\frac{dZ}{dt} = -\gamma Z.$$

We've seen before that $\frac{dZ}{dt} = \gamma Z$ gives exponential growth. When we add in the negative sign, it gives exponential decay. So the resulting function would look like this." Jessie drew Figure 9.1 on the whiteboard.

141

"But then when there are humans to attack, the zombies thrive the more contact they have with the humans. And they thrive in two ways, one by having something to eat, so it counteracts the starvation factor, and two by converting humans into additional zombies. Both of those factors are taken into account by having a term δHZ.

"So our equation for the rate of change in the number of zombies is

$$\frac{dZ}{dt} = -\gamma Z + \delta HZ."$$

"Why δ rather than β?" asked Ortiz. "If the equation for humans has a term that subtracts βHZ, shouldn't the zombie equation have a term that adds βHZ?"

"It's not that simple. When a zombie encounters a human, and the human loses, some of the time the human becomes a zombie and some of the time the human is killed and does not rise again. It's only the ones that become zombies that get added into the zombie population. Also, when a zombie obtains sustenance by eating a human, that helps the survival rate of the zombies as well. So we need different constants for the two different equations."

"Makes sense," said Ortiz.

"Okay, I said. "So we have the two equations

$$\frac{dH}{dt} = \alpha H - \beta HZ,$$

$$\frac{dZ}{dt} = -\gamma Z + \delta HZ."$$

"What does that tell us?"

"Well first let's clean them up a little by factoring the right-hand sides:

$$\frac{dH}{dt} = H(\alpha - \beta Z),$$

$$\frac{dZ}{dt} = Z(-\gamma + \delta H).$$

"Then notice that the critical points occur when $\frac{dH}{dt}$ and $\frac{dZ}{dt}$ are 0. So this is when either $H = Z = 0$ or when $H = \frac{\gamma}{\delta}$ and $Z = \frac{\alpha}{\beta}$."

"What are critical points again?" asked Marsha.

"They're points where the numbers for H and Z won't change as time progresses. So both of their derivatives are 0. They remain fixed."

"Why not $H = 0$ and $Z = \frac{\alpha}{\beta}$ or $Z = 0$ and $H = \frac{\gamma}{\delta}$?" asked Angus.

"That only makes one of the derivatives equal to 0. We need both equal to 0."

"But $H = Z = 0$ means both populations are gone," said Angus.

"Exactly. This is when the entire human population dies off, meaning that there's nothing left for the zombies to eat and they die off, too," said Jessie.

"So the bad news is we go extinct. But the good news is they do, too," I said.

"That'll teach them," said Angus.

"But that's only one possibility," said Jessie. "There is another critical point."

"And what about that one?" asked Ortiz.

"That's the other fixed point. In other words, if we had exactly $\frac{\gamma}{\delta}$ humans and exactly $\frac{\alpha}{\beta}$ zombies, then the numbers

143

would stay there. The new babies born to uninfected humans would exactly balance the loss of humans to zombies."

"Okay. So there is hope for the human race," said Marsha. "But what if the numbers are not exactly those values?"

"We can look at it in the phase plane," said Jessie.

"The what?" asked Angus.

"The HZ-plane. We let the y-axis be Z and the x-axis be H. Then, as time progresses, we map out the point in the HZ-plane corresponding to how many humans and zombies there are at that time."

"But how do we figure that out?"

(X. Predator-prey model continued on p. 184)

"We use the equations to draw the curves corresponding to time passing and we get a diagram like this" (Figure 9.2).

"But what does it mean for our chances of survival?" asked Angus.

"Suppose we start at a particular value on one of these curves," replied Jessie. "So there is a particular number of humans H and zombies Z. Then the differential equations tell us that we will stay on the curve, just travelling around it in the direction of the arrows."

"What does that tell us?" asked Ortiz.

"Look here." She pointed at Figure 9.3 which she had just drawn.

"When we start on the curve at point A, the number of zombies and the number of humans is large, so there are lots of 'interactions.' Plenty for zombies to eat, and the number of zombies grows while the number of humans shrinks. But as time passes, we move to point B where the number of zombies has grown and the number of humans has shrunk.

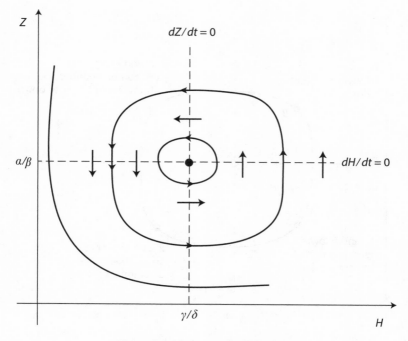

Figure 9.2: Curves in the HZ-plane.

Then the human population shrinks to such a point that the zombies can't find enough humans to feed upon or to convert, so the zombie population starts to drop. This takes us to point C. Then with fewer zombies, the human population has a chance to recover and we get to point D. But once the human population gets large enough, the zombies once again have plenty to feed on and to convert, and their number grows again, bringing us back to point A where we started. And the numbers cycle like that forever."

"What do you mean?" asked Ortiz. "The number of zombies goes up and down like a yo-yo forever?"

"Yup. It looks like this." She drew Figure 9.4.

145

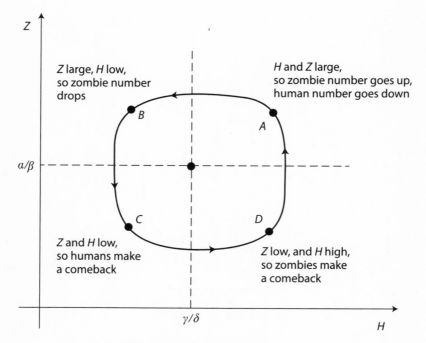

Figure 9.3: The predator-prey curve.

"So the good news is that humankind could survive this way," continued Jessie. "We could live in a world in which the zombies lived, too. The bad news is that in this model, we never get rid of the zombies. They keep making comebacks.

"But keep in mind this model doesn't take into consideration that humans are smart. In early human history, we were hunted by other beasts, and our population and theirs probably followed a cyclic oscillation like this. But eventually, we used our intelligence to outsmart the predators. We moved into caves and sharpened spears and used our brains to figure out how to beat them. And eventually, we wiped them out and became the dominant species. So if we can

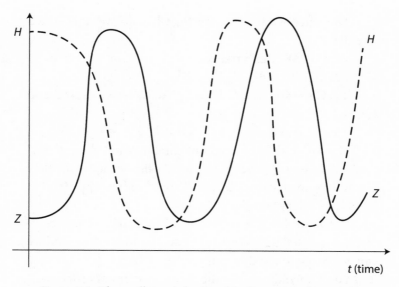

Figure 9.4: The oscillatory behavior of the human-zombie system.

survive the zombies long enough, the hope would be that the same would hold true for us. We just have to survive long enough."

"So how do we do that?" asked Thora softly. Everyone turned to look at her.

"We head north," I said.

"North?" said Sylvia. "Why would we do that? We're safe enough here."

"How much more candy do we have access to, Sylvia?" I asked. "We can't just stay here in the Science Center."

"Why north?" asked Marsha.

"Because they hibernate when it gets cold," I replied. "So the farther north we go, the more time they're hibernating. If we make it to winter, they'll be hibernating all the time. But we have to make it that long."

147

"And once we make it north, what do we do?" asked Marsha.

I smiled. "We build tree forts."

"What?"

"It's Connor's idea," I said. "Actually, Connor's and Cole's."

"What?" said Connor. Cole smiled.

"The two of them figured out zombies can't climb trees."

"But if we're up in some tree fort," said Sylvia, "won't they just wait at the base of the tree? Won't we be trapped? That seems like a very bad plan."

"Not if we build a collection of forts in a sequence of trees, connected by bridges. They surround one tree, and we can always go to another and get down there."

"You know it's exactly the way our ancestors survived," added Jessie.

"Maybe your ancestors," said Marsha.

"All of our ancestors," continued Jessie. "We're descended from the apes. And the apes survived in the trees, moving from one tree to another, and avoiding the predators on the ground. So the zombies have devolved, going back to an earlier evolutionary form. We devolve with them."

"Makes sense to me," said Angus.

"And how do we get north?" asked Ortiz.

"I've got some snowplows I think you'll like," I replied.

Ortiz shrugged. "I'm in," he said.

"Me, too," said Marsha.

"And me," said Thora. "When do we leave?"

'Now," I replied. "The sooner the better. Gunderson's waiting in the truck."

"Oscar's in the truck?" said Sylvia, smiling for the first time. She said it with an enthusiasm that made me wonder if Jessie wasn't the only one with whom Oscar had had a relationship.

"We'll all ride in the truck to the Town Garage," I continued. "There, we'll pick up the other snowplows. Then we'll stop at the Walmart on the way out of town to pick up the supplies we'll need."

"They have guns and ammo there," said Connor.

"Yes, they do," I said.

"What about gas or diesel or whatever the snowplows run on?" asked Marsha.

"They have fuel barrels at the town garage for times when the power's out, and they can't get it from a gas station. We'll load them into the back of one of the trucks. Should be enough to get us to Canada."

"I don't have a passport," said Marsha.

"If there's anyone there asking for passports, that'll be the happiest day of my life," I said. "But it's extremely unlikely."

"Let's get moving," said Jessie.

"All right. Connor, give your bat to Angus."

"No, Dad. Let Ellie give her bat to Angus. It's my bat."

"Connor, he's one of the stars of the baseball team," said Ellie.

"So give him your bat," said Connor.

"Connor," I said. "Hand it over."

He did so grudgingly.

"Thanks," said Angus to Connor. "I'll just borrow it."

"Okay," I said. "I'll take the lead and Angus the tail. We'll go single file down to the snowplow. I want all of you except Jessie, Gunderson, and me to get in the back of the truck. Then we'll drive over and pick up the other two trucks. Let's go."

I opened the door, checked the hall was clear, and headed down toward the stairwell, with Connor and the others

149

right behind me. We descended the stairs without incident, stepping carefully around bodies in the way. I pushed open the exterior door, and we filed out.

Suddenly Rusty came running up, barking. While I was considering the implications of this, I heard Sylvia say, "Oh, Oscar! It's so good to see you."

I turned around in time to see Gunderson with his arms outstretched, headed straight for Sylvia. She in turn headed straight for him, ready to receive his hug.

"Sylvia, no," I yelled, but it was too late She wrapped her arms around him, accepting his embrace. "Oh, Oscar. I have missed you." Her expression suddenly changed to surprise and then fear as Oscar tightened his grip.

"Oh, no, Oscar," she moaned, as he leaned in and took a huge bite of her neck. Blood sprayed out of her carotid artery. Gunderson looked pleased with himself, as blood poured down his jaw. Everyone stood frozen.

Gunderson let her body fall to the ground and turned to look hopefully at the rest of us. His glasses had disappeared and it appeared that he was having some trouble focusing on us. He let out a low moan and then charged forward, limp gone, and proceeded to slam face-first into a tree. Blood started to pour from his nose. Connor stifled a laugh.

"Everybody out of the way," I said. "This one's mine."

"No," said Angus. "I've got it."

"No," said Jessie. "Please. Let me."

She looked like she might cry.

Angus looked at me and I nodded. He handed Jessie his bat.

"Hey, Oscar," she said, "I need you to come over here."

At the sound of her voice, Gunderson turned, squinting in her direction. Then he charged. She wound up and swung, catching him right on the bridge of his nose. There was a cracking sound and then he fell backward, hitting the ground hard. He lay on the ground spasming for a moment and then lay still. Jessie slowly lowered the bat.

"I don't know if that was officially his skull," said Angus, "but something sure cracked."

I moved toward Jessie, but she waved me off.

"Let's get out of here," she said.

Epilogue

It's now been three months since the virus first appeared. It took us five difficult weeks, but we eventually reached Canada. Far enough north that although it's now August, it gets down to freezing at night. There are a few hours a day when the zombies are active, but we just make a point of hanging out in Tree City during those hours. Then, after it cools off, we climb down, take care of any zombies that have stumbled into our neighborhood, and we're good to go.

There aren't as many zombies around anyway. On our way up I-87, we passed a lot of zombies headed south. Looked like they were migrating to warmer climes, like retirees headed for the sun belt. Migration is another trait that must have very deep evolutionary roots. By October or November, we should be almost clear of them.

Our colony has grown in number to thirty-five. Tree City consists of twenty-one dwellings, all connected by an intricate bridgework system. Although we have many of the amenities of ground-based dwellings, we haven't yet figured out how to safely heat the town. Our most recent attempt burned two of the dwellings to the ground, and attracted a horde of zombies like moths to a lantern. That was not a good night.

Jessie and I share a tree house. So do Raphael and Marsha. Their's looks like a colonial two-story house, with windows that open, siding, and curtains. Thora, Angus, Ellie, Connor, and Cole built a tree fort that resembles a pirate ship. A big pirate ship, with five private cabins, a galley, and a living room. It is really quite amazing. In fact, they don't need five cabins anymore, since Thora and Angus have moved in together. I didn't see that one coming. Neither did Ellie, who had a bit of a crush on Angus. But Thora and Angus do make a striking pair. Thora carries two machetes, one strapped to each leg and, no big surprise, Angus packs a bat. Between the two of them, they are a formidable team.

Once enough of the zombies have departed for the south, we'll move into houses, with fireplaces, and/or generators and furnaces. Our hope is that by next spring, at least in this area, most of the zombies will be dead. Not enough human prey left to sustain them. Of course, zombies that migrate in one direction in the fall may very well migrate the other direction in the spring. We'll need to be prepared for that possibility.

Even if the number of zombies does drop dramatically, and we humans make a comeback, the Lotka-Volterra model implies that the zombies will then have plentiful prey and they will make their own comeback. We could be locked in this dance forever.

But the Lotka-Volterra model does not take into account that the prey might actually be intelligent, with the ability to modify its behavior. So, for instance, in our little group, we all wear protective gear. We travel in teams. We live a nocturnal lifestyle. And we exercise extreme caution. So far, we have only lost a few.

Unfortunately, zombies are not the only danger we face. We might need to add a new term to the equations. A term that takes into account the other survivors, and the fact that many of them are perfectly willing to kill for the shrinking pool of remaining resources. Turns out that some of the survivors are substantially more deadly than the zombies. I'm guessing that's been true throughout human history, that the danger from other humans almost always trumps the danger from the non-humans.

We've used calculus in those battles as well, to help protect ourselves from attack. But that's another story. One I don't have time to recount at this point.

I hope you found this story and the mathematics within it helpful. Calculus took humankind thousands of years to develop. But it is that ability to develop a system like calculus, and to use it to our advantage, that differentiates us from the zombies. Using our intelligence, we will eventually defeat them . . . if we don't kill each other first. I wish you the best of luck in your battle for survival.

–Craig Williams

appendix a

Continuing the Conversations

I. Growth (continued from p. 24)

"It's actually a little more complicated," I said.

"How so?" asked Angus.

"Figure it takes about fifteen minutes from when someone is bitten to when they become a zombie. That's how long it took for the one case I've seen. So if we consider the total number of zombies at discrete fifteen-minute intervals, and we assume that during the first fifteen minutes of infection an infected person will not infect anyone else, but after that they will infect two people every fifteen minutes, then we can get a recurrence relation that describes the situation."

"What's a recurrence relation?" asked Marsha.

"It's a discrete function that depends on its previous values."

"What's a discrete function?" asked Marsha.

"This is going to take a long time," said Gunderson.

"It just means a function that has values at discrete time intervals. In our case every fifteen minutes. So for instance, let $Z(n)$ be the number of zombies after n fifteen-minute time intervals have passed since the infection started. So $Z(0) = 1$, and $Z(1) = 1$, since that first zombie was incapable of infecting anyone else for the first fifteen minutes."

"And $Z(2) = 3$," said Angus, "because that first zombie could then infect two other people during the next fifteen minutes."

"Right," I said. "And then in general, at time $t = n$, the number of zombies will be the number at the end of the previous time interval, which is $Z(n - 1)$, plus the number of new zombies we added during the nth fifteen-minute time interval. And the new zombies are only produced by zombies that were around since two time intervals ago, so they produce $2Z(n - 2)$ new zombies."

"So then," said Angus, "the equation for the value of Z at time interval n is

$$Z(n) = Z(n - 1) + 2Z(n - 2)."$$

"Exactly. And that's our recurrence relation that describes the growth in the number of zombies."

"I hate to seem stupid, but how do you use that to find out the number of zombies? It's just a bunch of Z's and n's." said Marsha.

"We could write down the numbers one at a time," I replied. "Since $Z(0) = 1$ and $Z(1) = 1$ and $Z(2) = 3$, we can compute all the others. We get $Z(3) = 5$, and $Z(4) = 11$. So after one hour, which is four fifteen-minute time intervals, there were 11 zombies."

"That's not so many," said Angus. "I thought it would be worse."

"But if we want to know how many zombies there are now, after six hours, we don't want to keep writing down all the numbers of zombies for all of the 24 fifteen-minute time intervals," said Marsha. "That'll take forever."

"You're right. Instead, we can actually try to solve to find the function that satisfies this recurrence relation. We do that by guessing."

"You're allowed to guess in math?" said Marsha.

"Yup, so long as you check your guess and it works. Let's guess that the function that satisfies the equation is $Z(n) = s^n$ where s is some fixed number."

"Why that?"

"You'll see. It's an educated guess."

"Okay."

"So we plug it into the equation and we get

$$s^n = s^{n-1} + 2s^{n-2}.$$

Divide by s^{n-2} and we get

$$s^2 = s + 2.$$

This is

$$s^2 - s - 2 = 0.$$

So if we have a solution of the form s^n, the number s must satisfy $s^2 - s - 2 = 0$."

"I can factor that," said Marsha excitedly. "I remember that from high school."

"Amazing," said Gunderson sarcastically.

"Good, Marsha," I said. "What does it factor into?"

"I get $(s - 2)(s + 1) = 0$, so $s = 2$ and -1."

"Great," I said. "So there are two possibilities for s, 2 and -1. So $Z(n) = 2^n$ solves the equation and $Z(n) = (-1)^n$ also solves it. We'll use both."

"What do you mean 'use both'?" asked Angus.

"We'll take

$$Z(n) = A \cdot 2^n + B \cdot (-1)^n."$$

157

"Wait a minute. Where did A and B come from?" asked Angus.

"We want the most general solution. Notice that if 2^n solves the equation, so does $A \cdot 2^n$ where A is just a constant. We plug it in and get $A \cdot 2^n = A \cdot 2^{n-1} + 2A \cdot 2^{n-2}$. Factor out the $A \cdot 2^{n-2}$ and we have $2^2 = 2 + 2$, which is true."

"Okay," said Angus.

"Same holds for $B \cdot (-1)^n$. It's also a solution. We can add both of them together to get the most general solution:

$$Z(n) = A \cdot 2^n + B \cdot (-1)^n."$$

"But what are A and B?"

"Those are determined by the initial numbers. Remember we said that $Z(0) = 1$ and $Z(1) = 1$. Plugging those in gives us

$$1 = A + B,$$

$$1 = 2A - B.$$

Adding these gives us $2 = 3A$ so $A = 2/3$. Then since $A + B = 1$, we know $B = 1/3$.

"So our final solution is

$$Z(n) = \left(\frac{2}{3}\right) 2^n + \left(\frac{1}{3}\right) (-1)^n."$$

"That gives the number of zombies after n time intervals?" asked Marsha.

"Try it. When $n = 0$, we get 1. When $n = 1$, we get 1. When $n = 2$, we get $Z(2) = \left(\frac{2}{3}\right) 2^2 + \left(\frac{1}{3}\right) (-1)^2 = \frac{8}{3} + \frac{1}{3} = 3$.

When $n = 3$, we get $Z(3) = \left(\frac{2}{3}\right) 2^3 + \left(\frac{1}{3}\right)(-1)^3 = \frac{16}{3} + \frac{-1}{3} = 5$. It works!"

"Okay, but then how many zombies will there be after six hours?"

"Well, six hours is 24 fifteen-minute time intervals. So we just take $Z(24)$."

"What's that?"

"We'll have to use the calculator. I get

$$Z(24) = \left(\frac{2}{3}\right) 2^{24} + \left(\frac{1}{3}\right)(-1)^{24} = \frac{2^{25}}{3} + \frac{1}{3} = 11,184,811.$$

Around 11 million after six hours."

"Not good." said Jessie. "Not good at all."

II. Force (continued from p. 35)

(continued from p. 35)

"900 newtons is not so much," said Angus.

"That's just the average force," I continued. "If we want to know the maximum force when we hit the zombie, we can assume that the force occurs as a sine squared force."

"What does that mean?"

"Over the interval of .007 seconds, the force is least at the beginning and the end and most in the middle. So we use a sine squared function to model that. See here." I drew Figure A.1.

"The actual force experienced by the zombie's head varies over the time interval. It looks like a sine squared function that peaks in the middle. It would be given explicitly by $F = J \sin^2(\frac{\pi t}{0.007})$. So at time $t = 0$, there is not yet force. Then it grows to its peak value J at time $t = 0.0035$. Then it shrinks back to 0 at $t = 0.007$."

"But how high is it at the peak?" asked Angus.

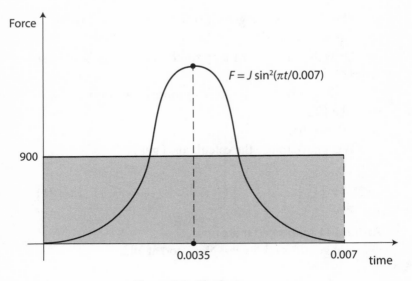

Figure A.1: The force curve.

"Well, we can think of it this way. The **impulse** is the amount of area under the force curve for the time interval of contact. We want the impulse for the average force we calculated before to be equal to the impulse for this new sine squared force. In the case of the average force, which we're thinking of as being constant, we get an area of 900 × .007 = 6.3 newton-seconds." I pointed to the shaded area in Figure A.1.

"We match the area under the sine squared curve with this area and that allows us to determine J."

"But to find the area under the sine squared curve, don't we have to integrate?" asked Angus.

"Exactly."

"But how do we find $\int J \sin^2 \left(\frac{\pi t}{0.007} \right) dt$?"

"We just use the double angle identity $\sin^2 \theta = \frac{1 - \cos(2\theta)}{2}$."

So

$$\int J \sin^2\left(\frac{\pi t}{0.007}\right) dt = \int J \left(\frac{1 - \cos 2(\frac{\pi t}{0.007})}{2}\right) dt$$

$$= J\left(\frac{1}{2}\int dt - \frac{1}{2}\int \cos\left(\frac{2\pi t}{0.007}\right) dt\right).$$

The first integral is easy, but to do the second integral, we need to use u-substitution. We let $u = \frac{2\pi t}{0.007}$. Then $du = \frac{2\pi}{0.007} dt$, so $dt = \frac{0.007}{2\pi} du$ and we get:

$$\int \cos\left(\frac{2\pi t}{0.007}\right) dt = \int \cos u \left(\frac{0.007}{2\pi}\right) du$$

$$= \frac{0.007}{2\pi} \sin u + C$$

$$= \frac{0.007}{2\pi} \sin\left(\frac{2\pi t}{0.007}\right) + C.$$

So we get

$$\int J \sin^2\left(\frac{\pi t}{0.007}\right) dt = J\left(\frac{t}{2} - \frac{0.007}{4\pi} \sin\left(\frac{2\pi t}{0.007}\right)\right) + C.$$

Then, for the definite integral,

$$\int_0^{0.007} J \sin^2\left(\frac{\pi t}{0.007}\right) dt = J\left(\frac{t}{2} - \frac{0.007}{4\pi} \sin\left(\frac{2\pi t}{0.007}\right)\right)\Big]_0^{0.007}$$

$$= J\left(\frac{0.007}{2}\right) = J(0.0035).$$

So the area under the sine squared curve is $0.0035J$. This then has to equal the area under the average force curve, which was 6.3 newton-seconds.

"So $6.3 = J(0.0035)$ and $J = 1800$ newtons. That's twice as high as the average force was. So the maximum force is actually 1800 newtons!"

III. The normal distribution (continued from p. 42)

"How would you show that the area under the curve $f(x) = \dfrac{1}{\sqrt{2\pi}} e^{\frac{-x^2}{2}}$ is 1?" asked Angus.

"Yeah, that's not so easy. Think of it this way. We need to show that the area under $g(x) = e^{\frac{-x^2}{2}}$ is $\sqrt{2\pi}$."

"Sure, because then when we divide that area by $\sqrt{2\pi}$, you get 1."

"Yup, and area under a curve is just an integral. So we need to show that $\displaystyle\int_{-\infty}^{\infty} e^{\frac{-x^2}{2}} \, dx = \sqrt{2\pi}$."

"But how do we do that integral? It looks hard. Can we use u-substitution?"

"Nope, doesn't work. In fact, none of the usual methods work."

"So, what do we do?"

"We use a clever trick." I smiled. I always liked this trick.

"Notice that $\int_{-\infty}^{\infty} e^{\frac{-x^2}{2}} \, dx$ and $\int_{-\infty}^{\infty} e^{\frac{-y^2}{2}} \, dy$ give the same value."

"Sure," said Angus. "By the time we integrate and evaluate at the limits of integration, the x and the y go away, so we'll get the same answer. The letter that we use doesn't matter."

"Okay, so let $I = \int_{-\infty}^{\infty} e^{\frac{-x^2}{2}} \, dx$, which we want to find.

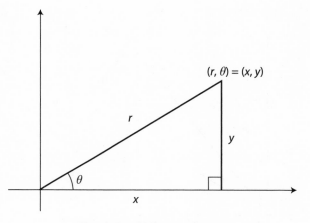

Figure A.2: Polar coordinates.

Then

$$I^2 = \left(\int_{-\infty}^{\infty} e^{\frac{-x^2}{2}} \, dx \right) \left(\int_{-\infty}^{\infty} e^{\frac{-y^2}{2}} \, dy \right)$$

$$= \int_{-\infty}^{\infty} \int_{-\infty}^{\infty} e^{\frac{-x^2}{2}} e^{\frac{-y^2}{2}} \, dy \, dx$$

$$= \int_{-\infty}^{\infty} \int_{-\infty}^{\infty} e^{\frac{-x^2-y^2}{2}} \, dy \, dx.\text{''}$$

"Okay..."

"Do you know polar coordinates?" I asked.

"Sure, when you replace the regular coordinates x and y by r and θ, where r is the distance to the origin, and θ is the angle between the line from the origin to (x, y) and the positive x-axis," he said as he drew Figure A.2. "Then $r^2 = x^2 + y^2$ by the Pythagorean Theorem."

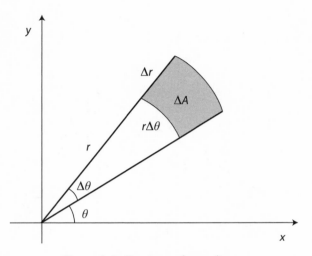

Figure A.3: Varying each coordinate.

"That's right."

"But something funny happens when you do a double integral in polar coordinates," said Angus.

"That's right," I said. "If we vary each of the coordinates a little, changing θ by $\Delta\theta$ and r by Δr, we get a region that looks like this." I drew Figure A.3.

"This region is very close to a rectangle. The bottom left arc has length $r\,\Delta\theta$ and the radial edge has length Δr."

"Huh? I don't see why that bottom left arc has length $r\,\Delta\theta$," said Angus.

"Think of it this way. That edge is actually part of a circle with radius r and angle $\Delta\theta$. The full circle would have circumference $2\pi r$, and the fraction of the circle that it gives is $\frac{\Delta\theta}{2\pi}$, so it should have length $\left(\frac{\Delta\theta}{2\pi}\right)2\pi r = r\,\Delta\theta$."

"Oh, yeah."

"Then, as Δr and $\Delta\theta$ get smaller, this region gets closer and closer to being a rectangle, so we know its area ΔA is approximately $r\,\Delta r\,\Delta\theta$."

"Yeah?"

"Yeah. Then, in the limit, instead of using $dA = dy\,dx$ in the integral, we replace it with $r\,dr\,d\theta$. So we get

$$I^2 = \int_{-\infty}^{\infty} \int_{-\infty}^{\infty} e^{\frac{-x^2-y^2}{2}}\,dy\,dx$$

$$= \int_{0}^{2\pi} \int_{0}^{\infty} e^{\frac{-r^2}{2}}\,r\,dr\,d\theta.$$

Now, we can do the inner r-integral by u-substitution, with $u = \dfrac{-r^2}{2}$. Then $du = -r\,dr$:

$$I^2 = \int_{0}^{2\pi} \int_{0}^{\infty} e^{\frac{-r^2}{2}}\,r\,dr\,d\theta$$

$$= \int_{0}^{2\pi} \int_{0}^{-\infty} e^{u}(-1)\,du\,d\theta$$

$$= \int_{0}^{2\pi} (-e^{u})\bigg]_{0}^{-\infty}\,d\theta$$

$$= \int_{0}^{2\pi} (0 - (-1))\,d\theta$$

$$= 2\pi.$$

So $I^2 = 2\pi$ and since I is clearly positive, $I = \sqrt{2\pi}$. That's what we wanted to show."

"That's a pretty cool trick," said Angus.

IV. The tangent vector (continued from p. 58)

"Why should this vector $\mathbf{v}(t)$ be tangent to the curve? And why should its length be the speed? Am I just slow?" asked Angus.

"No comment," said Gunderson.

"No, Angus, it's not obvious," I said. "Suppose we have this curve in the plane represented by $\mathbf{p}(t) = \langle f(t), g(t) \rangle$. So $f(t)$ gives the x-coordinate and $g(t)$ gives the y-coordinate of a point on the path. Let's think about the vector-valued function $\mathbf{v}(t) = \langle f'(t), g'(t) \rangle$.

"Since we have those derivatives in there, let's stick in the definition of the derivatives. Do you remember that?"

"The official definition?" said Angus. "Doesn't that have to do with a limit as Δt goes to 0?"

"That's right." I wrote it on the board:

$$f'(t) = \lim_{t \to 0} \frac{f(t + \Delta t) - f(t)}{\Delta t}.$$

"I don't know what you're talking about," said Marsha.

"It's actually just an expression that gives the slopes of the secant lines that approach the slope of the tangent line, which is what a derivative is." I drew Figure A.4.

"The expression $\frac{f(t+\Delta t) - f(t)}{\Delta t}$ is just the slope of the line passing through the two points $(t, f(t))$ and $(t + \Delta t, f(t + \Delta t))$. Then as we let Δt shrink to 0, we get the slope of the tangent line."

"If you say so, " said Marsha.

"I remember this," said Angus.

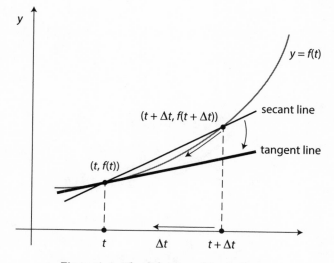

Figure A.4: The definition of the derivative.

"Then if we replace the two derivatives in $\mathbf{v}(t) = \langle f'(t), g'(t) \rangle$ with this expression," I continued, "we get

$$\langle f'(t), g'(t) \rangle =$$

$$\left\langle \lim_{\Delta t \to 0} \frac{f(t + \Delta t) - f(t)}{\Delta t}, \lim_{\Delta t \to 0} \frac{g(t + \Delta t) - g(t)}{\Delta t} \right\rangle$$

$$= \lim_{\Delta t \to 0} \frac{1}{\Delta t} (\langle f(t + \Delta t), g(t + \Delta t) \rangle - \langle f(t), g(t) \rangle)."$$

"Double trouble," said Angus.

"Just hang on a second," I said. "Take a look at that vector

$$\mathbf{w} = \langle f(t + \Delta t), g(t + \Delta t) \rangle - \langle f(t), g(t) \rangle."$$

I drew Figure A.5. "Notice that as Δt approaches 0, the direction of this vector approaches the tangent direction.

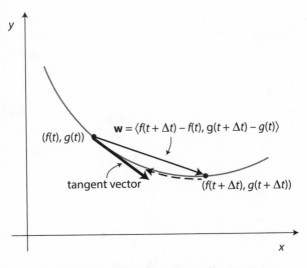

Figure A.5: The velocity vector is tangent.

The fact that it's multiplied by $\frac{1}{\Delta t}$ will only affect its length, not its direction."

"I see that," said Angus. "So in the limit, it's tangent to the curve. But what about its length? Since Δt is going to 0, and that's in the denominator, won't that make its length go to infinity?"

"Not actually. The vector **w** in the numerator is also getting shorter and shorter as Δt goes to 0, so those two effects counteract each other."

"So then what happens to the length?"

"The length of any vector $\mathbf{u} = \langle a, b \rangle$ is obtained by applying the Pythagorean Theorem. So its length is $\sqrt{a^2 + b^2}$. In this case, the length of the vector $\mathbf{v} = \langle f'(t), g'(t) \rangle$ is just $\sqrt{f'(t)^2 + g'(t)^2}$. But since $f'(t)$ is the speed with which the x-coordinate is changing and $g'(t)$ is the speed with which the y-coordinate is changing, $\sqrt{(f'(t)^2 + g'(t)^2}$ will be the

168

speed with which the point is moving along the curve. So, the velocity vector $\mathbf{v}(t) = \langle f'(t), g'(t) \rangle$ is a tangent vector to the curve with length exactly the speed that the point moves along the curve."

V. Pursuit (continued from p. 59)

"But how do you know that the path the zombie takes will be that curve?" asked Angus.

"Differential equations," I replied. "Let's start the zombie at the origin and the dean at the point $(a, 0)$ on the x-axis, a distance a away. Let's assume the dean is moving at constant speed of s_D ft./sec. in the vertical direction. So his coordinate at time t after he starts is $(a, s_D t)$."

"Makes sense," said Angus.

"Let's say the zombie has speed s_Z. We will let (x, y) be the position of the zombie at time t. So each of x and y depend on t. Since the tangent vector of the zombie always points toward where the dean is at that moment, we know that the tangent line to the zombie's path always hits the dean at that time." I drew Figure A.6.

"So we just take the line from the zombie's position $(x(t), y(t))$ at time t to the dean's position $(a, s_D t)$ at that same time," I continued.

"Okay," said Angus.

"So then, it's the tangent line to the zombie's curve. So its slope is $\frac{dy}{dx}$."

"Right."

"But since (x, y) and $(a, s_D t)$ are points on the line, its slope is also $\frac{s_D t - y}{a - x}$. So:

$$\frac{dy}{dx} = \frac{s_D t - y}{a - x}.$$

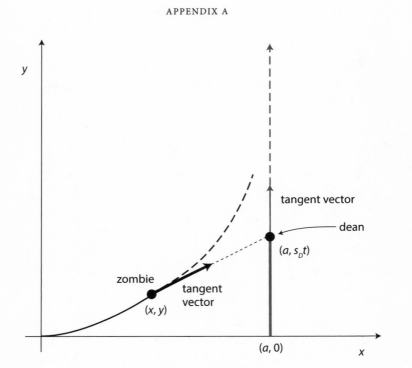

Figure A.6: Zombie chasing the dean.

That differential equation must then be satisfied by the zombie's curve."

"What does that tell us?" asked Jessie.

"Well, notice at time $t = 0$, both the zombie and the dean are on the x-axis. So the line between them at that time has slope 0. So $\frac{dy}{dx}$ equals 0 at time $t = 0$."

"Yes," said Jessie.

"But if the zombie catches the dean," I continued, "they will then have the same x-coordinate. That occurs when $x = a$."

"But when $x = a$, the denominator in the right side of that differential equation is 0, so $\frac{dy}{dx} = \infty$," said Angus.

170

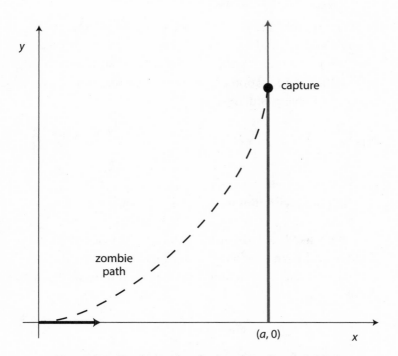

Figure A.7: Zombie catches the dean from directly behind.

"Yup," I agreed. "The zombie always catches up to the dean from directly behind. The zombie's curve must have a tangent vector that is horizontal at the beginning and vertical at the end." I drew Figure A.7.

"But then how do you figure out an explicit equation for the zombie's curve?" asked Jessie.

"That's hard," I replied. "If you want to see it, you should look at the book *Chases and Escapes* by Nahin" ([7]).

"I don't really have time for that right now," said Jessie, giving me a look that made it clear how ridiculous my suggestion was.

"I guess not," I said. "But turns out that if we let s_Z be the zombie's speed, and $r = s_D/s_Z$ be the ratio of their speeds, and we assume the zombie moves faster than the dean, so $r < 1$, then the equation of the curve looks like this." I wrote out the following equation:

$$y(x) = \frac{r}{1-r^2}a + \frac{1}{2}(a-x)\left[\frac{(1-x/a)^r}{1+r} - \frac{(1-x/a)^{-r}}{1-r}\right].$$

"Sorry I asked," said Jessie.

"It looks messy, but we can still tell a lot from it."

"Like what?"

"Well, like I said, the zombie catches the dean when they both have the same x-coordinate, which is when $x = a$. Then at that x-value, the second term on the right is 0, and we just have

$$y(a) = \frac{r}{1-r^2}a.$$

The rest of the equation disappears."

"That is a big improvement," said Jessie.

"Yup. For instance, if the zombie moves twice as fast as the dean, then $r = \frac{1}{2}$, and the zombie catches the dean at y-coordinate $\frac{\frac{1}{2}a}{1-(\frac{1}{2})^2} = \frac{2a}{3}$."

"Wow. The dean doesn't get far at all."

"Yes, but if the zombie only moves a little faster than the dean, say $r = \frac{9}{10}$, then the dean doesn't get caught until he has y-coordinate $\frac{90a}{19}$. As their speeds get closer and closer, the capture point gets farther and farther up in the y-direction."

"What happens if they have the same speed?"

"Then the dean never gets caught."

"I guess that's what you would expect."

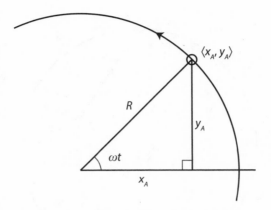

Figure A.8: Angus's path in circular pursuit.

VI. Circle pursuit (continued from p. 69)

"How do you figure out the differential equations that give the zombie's path?" asked Angus.

"Look at it this way," I said. "Angus, you were going in a circular path at a constant speed. So we can say that your path was given by

$$\langle x_A(t), y_A(t) \rangle = \langle R \cos(\omega t), R \sin(\omega t) \rangle$$

where R is your radius and ω determines how fast you ride around the circle."

"How'd you get that?" asked Angus.

"Look," I said as I drew Figure A.8.

"Since the angle is ωt, your x-coordinate is $R \cos(\omega t)$ and your y-coordinate is $R \sin(\omega t)$."

"Okay," said Angus.

"Then let $\langle x_Z(t), y_Z(t) \rangle$ be the position of the zombie at time t. Let s_Z be the speed of the zombie. All we know is that the zombie's velocity vector $\mathbf{v}_Z(t)$ always points toward you,

173

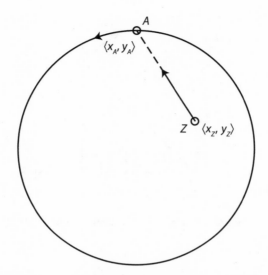

Figure A.9: Determining the zombie's path in circular pursuit.

Angus." I drew Figure A.9. "So at a given time t, we know that $\mathbf{v_Z}(t) = \langle x'_Z(t), y'_Z(t) \rangle$ points at $\langle x_A, y_A \rangle$."

"So take the vector that goes from $\langle x_Z, y_Z \rangle$ to $\langle x_A, y_A \rangle$. That's the vector $\langle x_A - x_Z, y_A - y_Z \rangle$. Divide it by its length to get a unit vector in the same direction. So we have

$$\left\langle \frac{x_A - x_Z}{\sqrt{(x_A - x_Z)^2 + (y_A - y_Z)^2}}, \frac{y_A - y_Z}{\sqrt{(x_A - x_Z)^2 + (y_A - y_Z)^2}} \right\rangle.$$

If we multiply this by s_Z, we have our velocity vector, since it has the right direction and it has the right length.

"Then we get

$$\mathbf{v_Z}(t) = \langle x'_Z, y'_Z \rangle =$$

$$s_Z \left\langle \frac{x_A - x_Z}{\sqrt{(x_A - x_Z)^2 + (y_A - y_Z)^2}}, \frac{y_A - y_Z}{\sqrt{(x_A - x_Z)^2 + (y_A - y_Z)^2}} \right\rangle.$$

"By considering each component separately, we obtain two coupled differential equations:

$$\frac{dx_Z}{dt} = s_Z \frac{x_A - x_Z}{\sqrt{(x_A - x_Z)^2 + (y_A - y_Z)^2}},$$

$$\frac{dy_Z}{dt} = s_Z \frac{y_A - y_Z}{\sqrt{(x_A - x_Z)^2 + (y_A - y_Z)^2}}.$$

"Adding in the expressions for x_A and y_A, we obtain:

$$\frac{dx_Z}{dt} = s_Z \frac{R\cos(\omega t) - x_Z}{\sqrt{(R\cos(\omega t) - x_Z)^2 + (R\sin(\omega t) - y_Z)^2}},$$

$$\frac{dy_Z}{dt} = s_Z \frac{R\sin(\omega t) - y_Z}{\sqrt{R\cos(\omega t) - x_Z)^2 + (R\sin(\omega t) - y_Z)^2}}.$$

"So these are the differential equations that need to be satisfied by the zombie's path. But they're too complicated to get an analytic solution. Instead, you get a computer to draw the path."

VII. Logistic growth (continued from p. 84)

"As Jessie explained," I said, "a disease often spreads so that the rate of spread is proportional to the number infected and the number not yet infected. So we have a differential equation that looks like this:

$$\frac{dZ}{dt} = kZ(P_0 - Z)$$

where P_0 is the initial population."

"But how do you find a function Z that is going to satisfy that?" asked Angus.

"We can use separation of variables," said Jessie.

"What's that?" asked Angus.

"You put all the terms with Z in them on one side and all the terms with t in them on the other side," said Jessie. "Like this."

$$\frac{dZ}{Z(P_0 - Z)} = kdt$$

"You're allowed to treat $\frac{dZ}{dt}$ like a fraction and multiply through by the denominator?"

"Yup."

"Okay, if you say so. You have the Ph.D. Then what?" asked Angus.

"Then you integrate both sides," continued Jessie.

$$\int \frac{dZ}{Z(P_0 - Z)} = \int kdt$$

"But how are you going to integrate $\int \frac{dZ}{Z(P_0-Z)}$? That looks hard."

"Not so hard," I said. "Just use partial fractions."

"Sounds vaguely familiar, but I don't remember that," said Angus.

"You just imagine writing the integrand $\frac{1}{Z(P-Z)}$ as a sum of two fractions. Like this."

$$\frac{1}{Z(P_0 - Z)} = \frac{A}{Z} + \frac{B}{P_0 - Z}$$

"But what are A and B?" asked Angus.

"We have to find them. Multiplying both sides of the equation by $Z(P_0 - Z)$, we get

$$1 = A(P_0 - Z) + BZ.$$

176

Since Z can be anything, we can set it to whatever we want. So we set it to 0, and we find $1 = AP_0$. So $A = 1/P_0$. Plugging that in, the equation becomes

$$1 = \frac{1}{P_0}(P_0 - Z) + BZ,$$

$$1 = 1 + Z\left(\frac{-1}{P_0} + B\right),$$

$$\frac{-1}{P_0} + B = 0,$$

$$B = \frac{1}{P_0}.\text{"}$$

"And so that means what?"

"That means we now know

$$\frac{1}{Z(P_0 - Z)} = \frac{1}{P_0 Z} + \frac{1}{P_0(P_0 - Z)}.$$

So, now we can integrate:

$$\int \frac{dZ}{Z(P_0 - Z)} = \int \frac{dZ}{P_0 Z} + \int \frac{dZ}{P_0(P_0 - Z)}.$$

"We can do both these integrals. We get

$$\int \frac{dZ}{Z(P_0 - Z)} = \frac{1}{P_0} \ln Z - \frac{1}{P_0} \ln(P_0 - Z) + C$$

$$= \frac{1}{P_0} \ln \frac{Z}{P_0 - Z} + C.$$

"Then $\int \frac{dZ}{Z(P_0-Z)} = \int k\,dt$ becomes

$$\frac{1}{P_0} \ln \frac{Z}{P_0 - Z} = kt + C,$$

$$\ln \frac{Z}{P_0 - Z} = P_0 kt + P_0 C,$$

$$\frac{Z}{P_0 - Z} = e^{P_0 kt + P_0 C}.$$

Notice $e^{P_0 C}$ is just a constant. Let's call it J. Then we have

$$\frac{Z}{P_0 - Z} = J e^{P_0 kt}.$$

Now solve for Z."

"How do we do that?" asked Angus.

"We multiply through by $P_0 - Z$."

$$Z = (P_0 - Z) J e^{P_0 kt}$$

$$Z = P_0 J e^{P_0 kt} - Z J e^{P_0 kt}$$

$$Z(1 + J e^{P_0 kt}) = P_0 J e^{P_0 kt}$$

$$Z = \frac{P_0 J e^{P_0 kt}}{1 + J e^{P_0 kt}}$$

"So that is our function that gives Z as a function of time t."

VIII. Average life span (continued from p. 103)

"How do you know that the average life span of a cell is $\frac{1}{\alpha}$?" I asked.

"I think I can explain that," said Jessie. "The claim is that any time you have a population P, be it cells or people or some other species or even radioactive particles, and that population dies off at a rate proportional to the size of the population, so $\frac{dP}{dt} = -\alpha P$, the average lifetime of an individual is $\frac{1}{\alpha}$."

"Right."

"So we'll think about it in terms of cells. Since P satisfies that differential equation, as we have already seen, P is given by $P = P_0 e^{-\alpha t}$."

"Right."

"But $\frac{dP}{dt}$ is really just a limit of $\frac{\Delta P}{\Delta t}$ where Δt is a short time interval starting at t, and ΔP is the change in P over that time interval."

"Sure."

"So this says that for short enough time intervals $[t, t + \Delta t]$, $\frac{\Delta P}{\Delta t} \approx -\alpha P$. Therefore, $\Delta P \approx -\alpha P \Delta t$. In other words, approximately $\alpha P \Delta t$ cells die in this time period, where the P we are using is P at the specific time t. Those cells that die in that short time interval have a life span of approximately t, since they died between time t and time $t + \Delta t$ and we are assuming Δt is short."

"Makes sense," I said.

"Okay, so now, we will add up all of the life spans of all of the cells in the population."

"You mean by integrating?" I asked.

"Yes, since integrating is essentially adding up. So we integrate the function $t \alpha P \Delta t$ and that gives us the sum of all of the life spans of all of the cells in the population."

"And the t came from the fact that this is approximately the life span of the cells that die in the time interval $[t, \Delta t]$?"

"Exactly."

"And the Δt becomes the differential dt at the end of the integral?"

"Yup. So the integral is $\int_0^\infty t\alpha P\, dt$."

"And the limits are 0 to ∞ because we are integrating over all possible life spans, from very short to very long?"

"Yes."

"And that gives the sum of all of the life spans of all of the cells?"

"Yes," nodded Jessie. "Then we divide that by the total number of cells we started with and that gives the average life span, which we call τ. So we get

$$\tau = \frac{1}{P_0} \int_0^\infty t\alpha P\, dt.$$

Now we substitute in the expression for P:

$$\tau = \frac{1}{P_0} \int_0^\infty t\alpha P_0 e^{-\alpha t}\, dt,$$

$$\tau = \alpha \int_0^\infty t e^{-\alpha t}\, dt."$$

"Okay."

"But I don't remember how to do that integral," said Jessie. "And I forget what you do when a limit of integration is infinity."

"We should be able to use integration by parts," I said.

"Oh yeah. Let's try that," said Jessie.

"We split the integral into two pieces, like deciding where to part your hair. I'll treat the integral as $\int u\, dv$. Then you can show $\int u\, dv = uv - \int v\, du$." (See p. 209 in Appendix B.)

"In this case, we'll set

$$u = t \qquad \text{and} \quad dv = e^{-\alpha t}dt.$$

Then $\quad du = \dfrac{du}{dt}dt = dt,$

and $\quad v = \displaystyle\int dv = \int e^{-\alpha t}dt = \dfrac{-1}{\alpha}e^{-\alpha t}.$

So according to the formula,

$$\tau = \alpha \int_0^\infty te^{-\alpha t}dt = \alpha \left(t\dfrac{-1}{\alpha}e^{-\alpha t} \right]_0^\infty - \int_0^\infty \dfrac{-1}{\alpha}e^{-\alpha t}dt \right)."$$

"But you still have an integral to do," said Jessie.

"Yes, but it's easier than the first one. We can do it directly, and we get

$$\tau = \alpha \left(t\dfrac{-1}{\alpha}e^{-\alpha t} - \dfrac{1}{\alpha^2}e^{-\alpha t} \right) \right]_0^\infty$$

$$= -\dfrac{t}{e^{\alpha t}} - \dfrac{1}{\alpha e^{\alpha t}} \right]_0^\infty$$

$$= \lim_{b \to \infty} -\dfrac{t}{e^{\alpha t}} - \dfrac{1}{\alpha e^{\alpha t}} \right]_0^b$$

$$= \lim_{b \to \infty} -\dfrac{b}{e^{\alpha b}} - \dfrac{1}{\alpha e^{\alpha b}} - \left(-\dfrac{1}{\alpha} \right)."$$

"Wait," said Jessie. "How do we take the limit of $-\dfrac{b}{e^{\alpha b}}$? It gives us $\dfrac{\infty}{\infty}$."

I smiled. "We use L'Hôpital's Rule."

"Why are you smiling?'"

181

"I just like it," I replied. "It says if we have a limit that is of the form $\frac{\infty}{\infty}$ or $\frac{0}{0}$, we can differentiate both the numerator and denominator and the limit will be the same.

"So in this case, since we get $\frac{\infty}{\infty}$, L'Hôpital's Rule holds. Since b is going to infinity, that is the variable we differentiate with respect to. So,

$$\lim_{b \to \infty} -\frac{b}{e^{\alpha b}} = \lim_{b \to \infty} -\frac{(b)'}{\left(e^{\alpha b}\right)'} = \lim_{b \to \infty} \frac{-1}{\alpha e^{\alpha b}} = 0.$$

Ultimately we get

$$\tau = \lim_{b \to \infty} -\frac{b}{e^{\alpha b}} - \frac{1}{\alpha e^{\alpha b}} + \frac{1}{\alpha} = 0 + 0 + \frac{1}{\alpha} = \frac{1}{\alpha}.$$

So the average life span of a cell is $\tau = \frac{1}{\alpha}$."

"That's it!" said Jessie.

IX. Equilibrium value (continued from p. 105)

"Why is the equilibrium value b^* for the number of uninfected brain cells equal to b_0/R_0?" I asked.

"Well, at the equilibrium values b^*, i^*, and v^*, the derivatives are all equal to 0. So we get these equations." She wrote them on the piece of paper.

$$\frac{db}{dt} = \lambda - \gamma b^* - \beta b^* v^* = 0$$

$$\frac{di}{dt} = \beta b^* v^* - \alpha i^* = 0$$

$$\frac{dv}{dt} = \kappa i^* - \mu v^* = 0$$

"From the third equation," she continued, "we know that $i^* = \frac{\mu}{\kappa}v^*$. Plugging that into the second equation gives us

$$\beta b^* v^* - \frac{\alpha\mu}{\kappa}v^* = 0,$$

$$v^*\left(\beta b^* - \frac{\alpha\mu}{\kappa}\right) = 0.$$

"Therefore, when the system is in equilibrium

$$b^* = \frac{\alpha\mu}{\beta\kappa}.$$

Plugging this value for b^* into the first equation, we have

$$\lambda - \gamma\left(\frac{\alpha\mu}{\beta\kappa}\right) - \beta\left(\frac{\alpha\mu}{\beta\kappa}\right)v^* = 0.$$

"So

$$v^* = \frac{\lambda\kappa}{\alpha\mu} - \frac{\gamma}{\beta}.$$

And finally, plugging this value for v^* into the third equation, we can solve to get

$$i^* = \frac{\lambda}{\alpha} - \frac{\mu\gamma}{\beta\kappa}.$$ "

"That's a lot of Greek letters."

"Really? You're a mathematician. I thought you loved Greek letters."

"I do. But I don't see what this all means."

"Well, if we substitute in with the basic reproductive rate $R_0 = \frac{\beta\lambda\kappa}{\alpha\mu\gamma}$, we can simplify the expressions a lot. For instance, substituting it into $b^* = \frac{\alpha\mu}{\beta\kappa}$, we get $b^* = \frac{b_0}{R_0}$ which is what we wanted to show."

"Oh, yeah. That's better."

"Doing the same for $v^* = \frac{\lambda\kappa}{\alpha\mu} - \frac{\gamma}{\beta}$, we get $v^* = (R_0 - 1)\frac{\gamma}{\beta}$. And $i^* = \frac{\lambda}{\alpha} - \frac{\mu\gamma}{\beta\kappa}$ can be rewritten as $i^* = (R_0 - 1)\frac{\gamma\mu}{\beta\kappa}$."

"That's a big improvement, " I said.

"Yeah, and notice that if R_0 is a lot bigger than 1, we can effectively ignore the -1 in each of these last two expressions. Then when we resubstitute back in with $R_0 = \frac{\beta\lambda\kappa}{\alpha\mu\gamma}$, they become

$$v^* \approx \frac{\lambda\alpha}{\kappa\mu} \qquad \text{and} \qquad i^* \approx \frac{\lambda}{\alpha}."$$

"Yes, and?"

Jessie shrugged. "So it's interesting they don't depend on β. The equilibrium numbers of these two don't depend on how infectious the virus is."

"That is strange."

"Yup and not only that. Look at i^*. Once we settle into this steady state, all that matters for the number of infected cells i^* is the rate of brain cell production in the hippocampus λ, and the death rate α of infected cells. None of the other parameters matter."

"Huh."

"But I guess for our purposes, the most important conclusion is that the number of uninfected brain cells settles down to a value that is the number of original brain cells b_0 divided by the basic reproductive rate R_0."

X. The Predator-prey model (continued from p. 144)

"From the two equations

$$\frac{dH}{dt} = H(\alpha - \beta Z),$$

$$\frac{dZ}{dt} = Z(-\gamma + \delta H),$$

we can figure out the slope field," said Jessie.

"The what field?" asked Angus.

"The slope field. Look, if we think of Z as a function of H, then $\frac{dZ}{dH}$ gives the slope of Z as a function of H."

"But all we know is $\frac{dZ}{dt}$ and $\frac{dH}{dt}$."

"Yes, but if we think of Z as a function of H, then by the Chain Rule, $\frac{dZ}{dt} = \frac{dZ}{dH} \frac{dH}{dt}$. So

$$\frac{dZ}{dH} = \frac{\dfrac{dZ}{dt}}{\dfrac{dH}{dt}}."$$

"Okay…" said Angus.

"So, dividing the one equation by the other," continued Jessie, "we get

$$\frac{dZ}{dH} = \frac{Z(-\gamma + \delta H)}{H(\alpha - \beta Z)}. \tag{A.0.1}$$

"Notice that when $H = \frac{\gamma}{\delta}$, the numerator equals 0, so $\frac{dZ}{dH} = 0$. So that holds along the entire vertical line $H = \frac{\gamma}{\delta}$. All of the slopes are horizontal there."

"I see that," said Angus.

"Also notice that when $Z = \frac{\alpha}{\beta}$, then the denominator yields 0, so then $\frac{dH}{dt} = 0$, so $\frac{dZ}{dH} = \infty$. So that's true for every point on the horizontal line $Z = \frac{\alpha}{\beta}$. There, all the slopes are vertical. So we know the slopes on these two lines, like in this picture." Jessie drew Figure A.10.

"Okay," said Angus.

"Also notice that when Z is near 0, $\frac{dZ}{dH}$ is near 0. So near the H-axis, the slopes are almost horizontal."

"Sure."

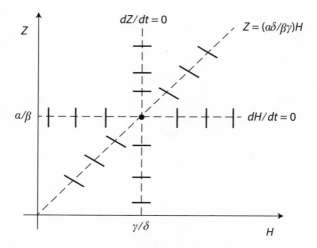

Figure A.10: The slope field in the HZ-plane.

"And when H is near 0, the denominator is near 0, so $\frac{dZ}{dH}$ is near ∞. So near the Z-axis, the slopes are almost vertical."

"Okay, but I'm still not sure what the point is," said Angus.

"Bear with me," said Jessie. "We can look at the diagonal line given by $Z = \frac{\frac{\alpha}{\beta}}{\frac{\gamma}{\delta}}H$ or $Z = \frac{\alpha\delta}{\beta\gamma}H$. If we plug that value in for Z and do the algebra, we get

$$\frac{dZ}{dH} = \frac{\frac{\alpha\delta}{\beta\gamma}H(-\gamma + \delta H)}{H(\alpha - \beta(\frac{\alpha\delta}{\beta\gamma}H))} = \frac{\frac{\alpha\delta}{\beta}(-1 + \frac{\delta}{\gamma}H)}{\alpha(1 - (\frac{\delta}{\gamma}H))} = \frac{-\delta}{\beta}.$$

So now we know all the slopes along the diagonal are $\frac{-\delta}{\beta}$."

"Okay," said Angus. "So we get all these slopes. But what do we do with them?"

"Then we can look at curves that appear to have those slopes. We get a picture like this." Jessie drew Figure A.11.

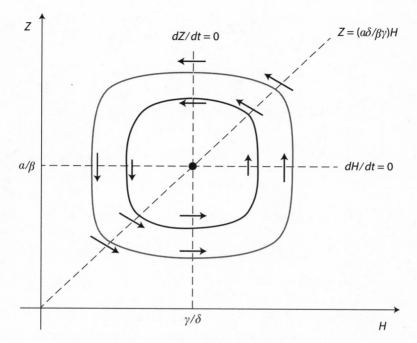

Figure A.11: Curves in the slope field in the HZ-plane.

"That's kind of cool. We just trace out curves that fit those slopes. But can we get an explicit expression for the curves?" asked Angus.

"I'm not sure," said Jessie.

"I bet we can," I said. "We can do a little algebra to separate the variables in Equation A.0.1, putting Z on the left and H on the right."

$$\frac{(\alpha - \beta Z)dZ}{Z} = \frac{(\delta H - \gamma)dH}{H}$$

$$\int \frac{(\alpha - \beta Z)dZ}{Z} = \int \frac{(\delta H - \gamma)dH}{H}$$

$$\int \left(\frac{\alpha}{Z} - \beta\right) dZ = \int \left(\delta - \frac{\gamma}{H}\right) dH$$

$$\alpha \ln(Z) - \beta Z = \delta H - \gamma \ln(H) + C$$

$$\alpha \ln(Z) + \gamma \ln(H) = \delta H + \beta Z + C$$

$$\ln(Z^\alpha) + \ln(H^\gamma) = \delta H + \beta Z + C$$

$$\ln(Z^\alpha H^\gamma) = \delta H + \beta Z + C$$

"We can also write this as

$$\ln(Z^\alpha H^\gamma) - \delta H - \beta Z = C."$$

"Can we solve that for Z as a function of H or H as a function of Z?"

"Unfortunately not. That has no explicit solution. It's what we call a transcendental equation. Both variables are defined implicitly in terms of one another."

"You mean like transcendental meditation?" asked Marsha.

"Not quite, but it's the same idea. The word 'transcendental' means not experienced, but knowable. In the context of mathematics, it means an equation that is not algebraic. In this case, it is not solvable for one variable in terms of the other."

"So, what do we do?" asked Angus.

"Well, the important fact is that there is an equation that is satisfied by Z and H of the form $F(H, Z) = C$. Then we call C a *constant of motion*."

"What does that mean?"

"It means that if we know the values of H and Z at some time, say they are initially H_0 and Z_0, then that gives us a value for C. Then for the rest of time, however those two change, they still have to satisfy the same equation $F(H, Z) = C$. So all of the values for all time lie on that curve $F(H, Z) = C$."

"I'm lost," said Marsha.

"Me, too," said Angus.

"Me, too," said Ortiz.

"Here's another way to look at it," I said. "Consider this function

$$F(H, Z) = \ln(Z^{\alpha} H^{\gamma}) - \delta H - \beta Z.$$

It's a function that gives a value for each point in the HZ-plane."

"Okay," said Angus. "So if I give you a certain value for the number of humans and number of zombies at a given time, you can give me the value of that function."

"Right. So now we could graph that function in 3-space, above the HZ-plane. And it might look like a mountain."

"Okay . . . ," said Angus.

"And if we intersected that mountain with horizontal planes at various heights, we would get contour curves."

"Sure."

"And we could draw those contour curves in the HZ-plane, just like we do with a contour map to help visualize a mountain."

"Yup."

"And the highest or lowest point in the mountain would just have a contour curve that is a point."

"Sure."

"Well, that is the point $(H, Z) = \left(\frac{\gamma}{\delta}, \frac{\beta}{\alpha}\right)$. The other curves come from intersecting horizontal planes at various heights with the mountain. And that's how we get the picture." I motioned to Figure A.11.

appendix b

A Brief Review of Calculus as Explained to Connor by Ellie

"What's all that calculus stuff Dad's always talking about?" asked Connor.

"It's a kind of mathematics," replied Ellie.

"What kind of mathematics?"

"Forget it. You wouldn't understand."

"You don't know, do you?"

"Connor, of course I know. I took a whole year of calculus in high school."

"You make it sound like you're not in high school anymore."

"Connor, I'm not. There won't be high school anymore. School is over for a very long time."

"I know that," said Connor. "And that's okay with me. So what's calculus?"

"It's about change. It's about measuring change. You know, how fast something changes."

"You mean like when somebody changes from human into zombie?"

"Well, yeah. It could measure that. It could tell you how fast the number of virus particles is growing in their body for instance. Or it could measure how fast a car is moving."

"We already have speedometers for that."

"Yes, but that speedometer is really telling you the value of the derivative for the car."

"Is that what calculus is? The derivative?"

"It's part of calculus, a big part of calculus. The other big part is called the integral. Those two things, the derivative and the integral, that's calculus."

"But why stick them together? You can't just take two random things and then say, hey, I invented a new course called calculus and it's these two things."

"No, of course you can't. It's because they're related. You know Isaac Newton, the guy who invented gravity?"

"He didn't invent gravity," said Connor. "He discovered gravity."

"Whatever. Anyway, he and another guy, Liebniz, they figured out that the derivative and the integral are actually opposites of one another."

"What do you mean, opposites?"

"One cancels out the other. You do one and then you do the other and you're back where you started. So that's why both are in calculus."

"Okay, so explain them to me. What's a derivative?"

"Are you kidding? It took me a year to learn calculus. Do you really expect me to explain it all to you?"

"Just give me the basics. Enough so when Dad starts blabbing about it the next time, I'll know some of what he's talking about. What's a derivative?"

"Look, I have lots of stuff I'm supposed to be doing. Dad wanted me to separate the different kinds of bullets into different baggies, so we don't mix them up anymore."

"It wasn't my fault they got mixed up."

"I didn't say it was. But Dad asked me to straighten it out."

"So just tell me about calculus while you're doing it. I'll even help you." Connor grabbed the box of baggies off the shelf and placed the bowl of bullets next to it on the table.

"Okay," said Ellie, as she sat down and pulled out a handful of baggies. Connor sat down and pulled out several more. They began to sort the bullets.

"Let's start with the derivative," said Ellie."I'll give you an example. Suppose you're running from a zombie, and you start at the doors of the Science Center and you run in a straight line toward the Lattimer Performing Arts Center, which is 70 feet away. Let's say $f(t)$ is the number of feet you've run after t seconds. So if after five seconds you've run 20 feet, then $f(5) = 20$. If after 10 seconds, you've run 45 feet, then $f(10) = 45$. Do you get it?"

"Yeah, I get it."

"Okay. Then you probably noticed that your speed is changing. You ran farther in the second five seconds than you did in the first five seconds. That's just because you're getting up to speed."

"Sure," said Connor."I noticed that. I run faster once I get going."

"Good, so look here." Ellie put down the bullets in her hand, and took a piece of paper and a pencil off the shelf. She drew Figure B.1.

"Here's a graph of where you are in terms of the time. It shows how far you've run on the vertical axis for any given time on the horizontal axis. That graph has to pass through $(5, 20)$ and $(10, 45)$, since we know $f(5) = 20$ and $f(10) = 45$."

"I think I could run a lot faster than that," said Connor.

"Okay, first, that is irrelevant, and second, no you couldn't."

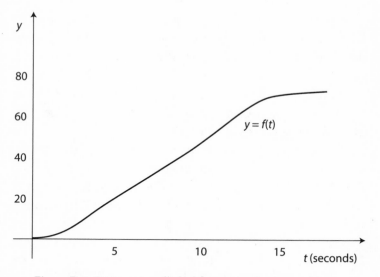

Figure B.1: Distance travelled while running from a zombie as a function of time.

"People run the hundred yard dash in under 10 seconds."

"Yeah. Those are professional world-class athletes. Now, do you want to know about derivatives or not?"

"Okay, keep going."

"Look at the graph. At first, the graph is not that steep near $t = 0$. That's because you're just getting started, and you're not running that fast yet. Then you get going faster and the graph gets steeper. Then, at the end, you slow down, when you reach the doors of Lattimer.

"You measure how fast you're going at any given instant t_0 by considering the point on the graph corresponding to $t = t_0$, and taking the line through that point that's tilted at the same angle as the graph is tilted at that point." She drew Figure B.2.

"That's called the tangent line. How steep that line is measures how steep the graph is at that point, which tells you

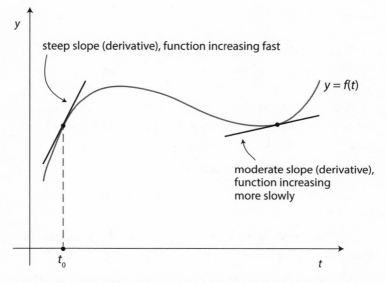

Figure B.2: The slope of the tangent line to a graph measures how steep the graph is at that point.

how fast your function is changing at that point. We measure how steep the line is by its slope."

"I already know about slopes of lines."

"Good for you. When you're going slowly, the slope of the tangent line isn't steep. But when you're going your fastest, the slope is the steepest. So that's all there is to a derivative $f'(t)$ of a function $f(t)$. It's just the slope of the tangent line to the graph of the function $f(t)$ at the point corresponding to a particular t."

"The slope of the tangent line. Got it."

"Yup, and the variable, which in this example was t, corresponding to time, could be anything. Lots of times, people use an x. So we'll have a function $f(x)$ with values that depend on x. Then its derivative $f'(x)$ also depends on x, since the slope of the tangent line varies as we change the values of x. The derivative of a function $y = f(x)$ can be written $f'(x)$, $\frac{df}{dx}$, y', $\frac{dy}{dx}$, or $\dot{y}$."

"I think I'd use $\dot{y}$. That's the shortest."

"That's what the physics people use. So maybe you'd be a physicist."

"I don't know what that is either, but let's stick to calculus for now. So you have this derivative, which is the slope of the tangent line. So who cares? What good is it?

What does the derivative tell us?"

"We can use the derivative to figure out a bunch of stuff. For instance,

if the derivative is positive somewhere, the function is increasing there.

This is pretty obvious. If the derivative is positive, then the slope of the tangent line is positive. So the tangent line is tilted up. So the function, which hugs the tangent line, must also be tilted up, like here" (Figure B.3).

"**If the derivative is negative somewhere, the function is decreasing there.**

Again, obvious. If the derivative is negative, then the slope of the tangent line is negative. So the tangent line is tilted down. So the function, which hugs the tangent line, must also be tilted down, like here" (Figure B.4).

"That makes sense," said Connor.

"Okay, now here's the important point.

If your function has a local maximum or minimum, the derivative has to be 0 there."

"What do you mean a local maximum or minimum?"

"You know, a place where it is the highest point or lowest point, at least for all nearby points, like here" (Figure B.5).

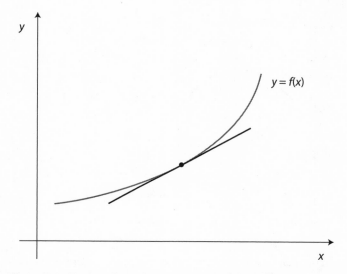

Figure B.3: Positive derivative means positive slope, which means the function is increasing.

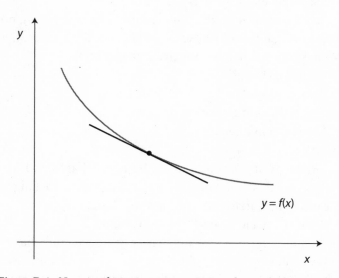

Figure B.4: Negative derivative means negative slope, which means the function is decreasing.

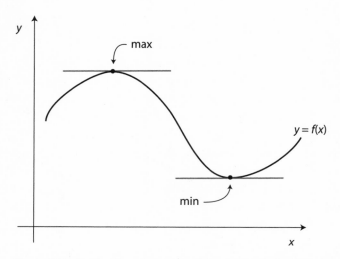

Figure B.5: At a local maximum or minimum, the derivative is 0.

"So what about the derivative there?"

"At a max or min, the function is neither increasing nor decreasing. So the derivative is neither positive nor negative. What number is neither positive or negative?"

"Is this a riddle?"

"No, it's a question."

"It's a stupid question."

"So you don't know?"

"It's zero."

"Right. So at a max or min, the derivative has to be 0. You can also see this in the picture. At a max or min, the tangent line has to be horizontal with slope 0" (Figure B.5.).

"Why is that a big deal?" asked Connor.

"Being able to figure out where your function has its highest value or lowest value is a big deal. In fact, this used to be the most common use of the derivative. Suppose you're the CEO of a multinational company that makes Ferraris."

"The company that makes Ferraris is Italian and it's called Ferrari."

"Whatever. It doesn't exist anymore anyway. But if you were the CEO and you had a function that represented the company's potential profit as a function of a bunch of variables, you would want to find the place where that function is maximized. You could find it by setting the derivative equal to 0, since at a highest point, the tangent line is horizontal with slope 0. Then of the points that you find with derivative 0, take the one with the highest value. Nowadays, that's not the relevant application anymore, since nobody's making Ferraris, but being able to find max and min is still really important."

"Okay, so it's useful. But it's only useful if you can find it.

How do you find the derivative?"

"Oh, God. This could take all day."

"Just give me the highlights."

"Okay, look. We need to be able to find the slope of the tangent line to the graph of a function $f(x)$ at any particular point x. The problem is that it's not even obvious how to define the tangent line. It's a line that hits the curve at exactly one point near the point of interest. But lots of lines do that. We want the one that hits the curve at the one point and that hugs the curve near there."

"What do you mean 'hugs the curve'?"

"I'll show you. What we do is take a line that hits the curve in two points, called a *secant line*. It hits the curve at the original point $(x, f(x))$ and at a nearby point $(x + \Delta x, f(x + \Delta x))$. Think of Δx as a really small amount like this" (Figure B.6).

"That picture's a mess."

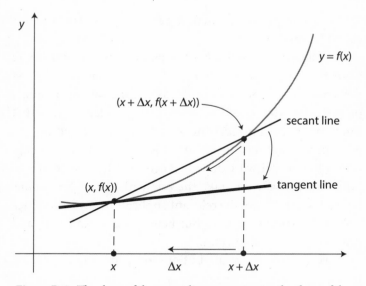

Figure B.6: The slope of the secant line approximates the slope of the tangent line.

"No, it's not. Do you see the secant line hitting the graph at $(x, f(x))$ and $(x + \Delta x, f(x + \Delta x))$?"

"Yeah, I see it."

"And can you see that if I shrink Δx smaller and smaller, the point $(x+\Delta x, f(x+\Delta x))$ moves down the curve toward the point $(x, f(x))$? And the secant line swings down and becomes the tangent line?"

"Yeah, I guess."

"So then the slope of the secant line would get closer and closer to the slope of the tangent line we want to compute. The slope of the secant line is

$$m = \frac{\Delta y}{\Delta x} = \frac{f(x + \Delta x) - f(x)}{(x + \Delta x) - x} = \frac{f(x + \Delta x) - f(x)}{\Delta x}."$$

"I know how to find a slope."

"Good, then you shouldn't look confused."

"I'm not confused."

"Okay, then maybe you just look it."

"I get it. That's the slope of the secant line and when Δx shrinks to 0, it becomes the slope of the tangent line."

"Right, so what we do is consider that slope m as Δx shrinks to 0. So this is our official definition of the derivative, where $\lim_{\Delta x \to 0}$ just means we see what happens as we shrink Δx to 0.

$$f'(x) = \lim_{\Delta x \to 0} \frac{f(x + \Delta x) - f(x)}{\Delta x}$$

"I know it looks messy," continued Ellie, "but from it, we can figure out a bunch of rules that make it easy to find derivatives."

"Like what?"

"Like this."

I. The Constant Rule. If c is a constant, $\dfrac{dc}{dx} = 0$.

"This just says that if $f(x) = 3$, for instance, $f'(x) = 0$ for all x. This isn't a big surprise, since the graph of $f(x) = 3$ is the horizontal line $y = 3$. Obviously, all of the tangent lines are horizontal with slope 0," so the derivative is always 0 (Figure B.7).

"Well, yeah. That's obvious."

"I'm glad you think so. Here's another one."

II. The Sum Rule. $\dfrac{d}{dx}(f + g) = \dfrac{df}{dx} + \dfrac{dg}{dx}$.

"Sounds okay."

"Then there's this one."

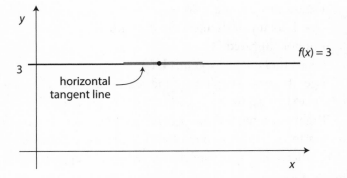

Figure B.7: For a constant function, slope and hence derivative is 0 everywhere.

III. The Constant Multiple Rule. $\dfrac{d(cf)}{dx} = c\dfrac{df}{dx}.$

"So constants pass right through the derivative," said Ellie.

"If you say so."

"You're not getting this, are you?"

"I get it. The derivative of 3 times f is 3 times the derivative of f."

"Okay, then if you're so smart, let's try you with a harder rule."

IV. The Power Rule. $\dfrac{d}{dx}(x^n) = nx^{n-1}.$

"What's the n?" asked Connor.

"It can be any number you want. So, for instance, $\dfrac{d}{dx}(x^2) = 2x^1$ and $\dfrac{d}{dx}(x^{100}) = 100x^{99}$."

"That doesn't seem hard."

"Good. So then if we want the derivative of $f(x) = 7x^3 + 4x^2 - 2x + 9$, the Sum Rule lets us take the derivative of each of the individual terms, the Constant Multiple Rule lets us pass the constants 7, 4, and 2 through the individual derivatives, the Power Rule lets us take the derivatives of each of the powers of x, and the Constant Rule tells us the derivative of 9 is 0. So we get $f'(x) = 21x^2 + 8x - 2$."

"That's cool. Let me do one."

"Okay, $f(x) = 4x^3 + 7x^2 - 2x + 8$."

Connor smiled. "Then $f'(x) = 12x^2 + 14x - 2$."

"Good. So now you can take a basic derivative. That's something no zombie could do."

"Oh thanks, so you're saying I'm smarter than a zombie."

"I wouldn't go that far. Hey, they could have a TV show called *Are You Smarter than a Zombie?*"

"Only if there were TV shows anymore."

"Good point. Okay, ready for the next rule?"

"There's more?"

"Yup. The next one is when you have two functions multiplied together and you want to take their derivative."

V. The Product Rule. $\dfrac{d}{dx}(f \cdot g) = \dfrac{df}{dx}g + f\dfrac{dg}{dx}.$

"There's also another one."

VI. The Quotient Rule. $\dfrac{d}{dx}\left(\dfrac{f}{g}\right) = \dfrac{f'g - fg'}{g^2}.$

"That's weird."

"And then there's this one."

VII. The Chain Rule. $\dfrac{d}{dx}(f(g(x))) = f'(g(x))g'(x).$

"Why's it called the Chain Rule? I don't see any chain."

"It's for when you have one function inside another, which is called a composition. I guess that's considered a chain of functions."

"You're making that up."

"Do you want to learn calculus or not?"

"Okay, okay. Keep going."

"All right. This next one is even weirder."

VIII. $\dfrac{d}{dx}(e^x) = e^x$.

"Okay, first, what is the 'e'?"

"You don't know 'e'? It's one of the most famous numbers ever invented!"

"Oh, you mean the number 'e.' I know that. It's 2.71828."

"I'm amazed you know it to that many decimal places."

"Jeff knows it to twenty decimal places. Or he did before he became a zombie."

"Well, that's what the e is. It's that number."

"So if you take the derivative of e^x, you get e^x back again? That's kind of crazy. It's its own derivative."

"Yup. In fact, that's really the reason why e is so famous, because of that."

"Oh."

"Okay, so now we can take a lot of derivatives. For example, $\frac{d}{dx}e^{kx} = ke^{kx}$."

"Why is that?"

"We just use the last two rules. We can think of e^{kx} as a composition of $f(x) = e^x$ and $g(x) = kx$. So $f(g(x)) = e^{kx}$. Then since $f'(x) = e^x$ again, $f'(g(x)) = e^{kx}, g'(x) = k$, so $\frac{d}{dx}(f(g(x)) = e^{kx}k$."

"I get that."

"Okay, then. Let's go for an even stranger rule."

IX. $\dfrac{d}{dx}\ln(x) = \dfrac{1}{x}$.

"Yeah, well I don't even know what $\ln x$ is, so let's skip that one."

"Fair enough."

"So is that it for derivatives?"

204

"No. It's not like you can learn all of calculus in fifteen minutes."

"So what else is there?"

"We can take derivatives again."

"What do you mean?"

"When we take the derivative of a function $f(x)$ and we get $f'(x)$, that's also a function of x. So there's nothing stopping us from taking its derivative, which we write as $f''(x)$ or $\frac{d^2 f}{dx^2}$. The **second derivative** is telling us how fast the derivative is changing."

"But what does that tell us about the graph of the original function?"

"It tells us about the concavity.

If $f''(x)$ is positive for some x, the graph is concave up there.

If $f''(x)$ is negative for some x, the graph is concave down there.

"At points where $f''(x)$ changes sign, we have points on the graph where the function changes concavity. These are so-called *inflection points*." Ellie drew Figure B.8.

"How do you remember what concave up and concave down mean?" asked Connor.

"I just remember it as concave up looks like a cup facing up," replied Ellie. "That's the way it curves. And concave down looks like a cup facing down."

"Okay."

"The second derivative is especially interesting when we have a function $f(t)$ that gives our position in terms of time t," continued Ellie. "Then $f'(t)$ is the velocity and $f''(t)$ is the acceleration function."

"Okay, enough," interrupted Connor. "I get derivatives. Now what about those things you called integrals?"

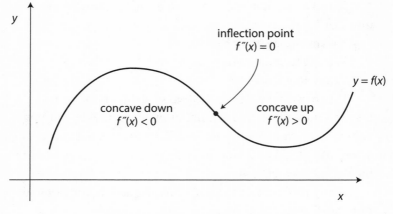

Figure B.8: $f''(x)$ determines concavity.

"Integrals?

They're the other half of calculus. Integration is the inverse operation of differentiation."

"Wait a minute. You never said anything about differentiation. What's that?"

"What do you mean? I just told you about differentiation."

"No you didn't. You told me about derivatives."

"Oh, my God," said Ellie. "Differentiation is just the act of taking a derivative."

"Then it should be called derivativization."

"Yeah, right," said Ellie, rolling her eyes. "Good thing you're not in charge of the dictionary."

"You know I'm right."

"I'm just going to ignore your suggestion. As I was saying, in taking a derivative, which is called *differentiation*, we start with a function $f(x)$ and we find a new function

$f'(x)$ which is its rate of change. In integration, we start with a function $f(x)$, and take its integral, written $\int f(x)dx$, which is a function such that its derivative is the original $f(x)$. In other words,

$$\frac{d}{dx} \int f(x)\, dx = f(x).$$

The two operations of integration and differentiation cancel each other out."

"What's that 'dx' at the end?"

"That's called the differential. You always put it there in an integral."

"Seems like a waste. Why not just have everybody agree to leave it out, and then we all save the time it takes to always put it in?"

"Good grief. It's there to tell you what variable to integrate."

"But there is only one variable, x."

"Yeah, but in the real world, there will often be more than one. So you need to know which one you are integrating."

"Okay. So let me try it. Since $\frac{d}{dx}x = 1$, then $\int 1\, dx = x$."

"You're sort of right."

"You can't be sort of right. You're either right or you're wrong. Which is it?"

"You're right that x is a function that is the integral of 1, since its derivative is 1. But there are a whole bunch of functions, the derivatives of which are 1. Take $x + 3$ for instance. Or $x + 7$. So, in fact, the most general function with its derivative equal to 1 is $x + C$ where C is any constant. So we write $\int 1 dx = x + C$, where C is called the *arbitrary constant*."

"That seems silly to always write a '+C' but okay, I get that."

"And just like there's a bunch of rules for doing derivatives, there's a bunch of **rules for doing integration**.

I. The Constant Rule. If k is a constant, $\int k \, dx = kx + C$.

II. The Sum Rule. $\int (f(x) + g(x))dx = \int f(x)dx + \int g(x)dx$.

III. The Constant Multiple Rule. $\int cf(x)dx = c \int f(x)dx$.

IV. The Power Rule. $\int x^n dx = \frac{1}{n+1}x^{n+1} + C$."

"That makes sense. You just reverse the Power Rule for derivatives."

"Exactly."

"And I bet I know the integral of e^x. It's got to be:

V. $\int e^x \, dx = e^x + C$."

"I guess Dad's right."

"What?"

"That you're smarter than you look."

"He never said that."

"You'll never know if he did or not. But you'll wonder for years and years."

"No I won't. What else do you got?"

"**VI.** $\int \frac{1}{x} \, dx = \ln |x| + C$."

"I told you. I don't care about the ln x."

"Whatever. Then let's talk about doing harder integrals, like $\int (x^2 + 1)^3 x \, dx$."

"It's a product so can you do a product rule?"

"Nope. Doesn't work. There is no regular product rule for integration."

"So what do you do?"

"You need u-**substitution**."

"What's that?"

"It says

$$\int f(g(x))g'(x)dx = \int f(u)du.$$

You let $u = g(x)$ and then $du = g'(x)dx$. Then if you're lucky, you can do the new integral."

"But how do we know if we have the right kind of integral for this?"

"We try it. So, for example, if we want to do $\int (x^2 - 1)^3 x\, dx$, we let $u = x^2 - 1$, the function inside the power. Then $du = \frac{du}{dx}dx = 2xdx$. So $xdx = du/2$, and we can do the substitution.

$$\int (x^2 - 1)^3 x\, dx = \int u^3 \frac{1}{2}\, du = \frac{1}{8}u^4 + C$$

$$= \frac{1}{8}(x^2 - 1)^4 + C."$$

"Got it."

"I doubt if you do, but let's keep going. I don't have all day."

"What's next?"

"Ummmm, let's do

Integration by Parts.

It says $\int u\, dv = uv - \int v\, du$.

"Notice that dv just means $\frac{dv}{dt}dt$ and du means $\frac{du}{dt}dt$. Then this is really just the Product Rule for differentiation in reverse. Remember, $\frac{d(uv)}{dt} = u\frac{dv}{dt} + v\frac{du}{dt}$. So $u\frac{dv}{dt} = \frac{d(uv)}{dt} - v\frac{du}{dt}$.

"Then integrate both sides:

$$\int u\frac{dv}{dt}\, dt = \int \frac{d(uv)}{dt}\, dt - \int v\frac{du}{dt}\, dt.$$

"Using the fact $dv = \frac{dv}{dt}dt$ and $du = \frac{du}{dt}dt$, and that differentiation and integration are inverse operations, we get

$$\int u\, dv = uv - \int v\, du.$$

"The trick to integration by parts is to pick each of u and dv in the original integral, so that the new integral is easier to do than the original."

"Yeah, but how do you know which is which?"

"Practice. For example, if we want to find $\int xe^x\, dx$, we could let $u = x$ and $dv = e^x\, dx$. Then $du = 1dx$ and $v = \int dv = \int e^x\, dx = e^x$."

"Don't you need a '$+C$' there?"

"No. There'll be a '$+C$' at the end, and that's enough. Then

$$\int \underbrace{x}_{u}\underbrace{e^x\, dx}_{dv} = \underbrace{x}_{u}\underbrace{e^x}_{v} - \int \underbrace{e^x}_{v}\underbrace{dx}_{du} = xe^x - e^x + C."$$

"Okay, I guess. So is that it? Do I know everything about integration now?"

"Are you kidding? We haven't even done definite integrals yet."

"What's that?"

"That's integrals with

Limits of Integration.

Suppose $F(x) = \int f(x)\, dx$. Then when we put numbers at the top and bottom of the integral sign, it means

$$\int_a^b f(x)\, dx = F(x)]_a^b = F(b) - F(a).$$

"$\int_a^b f(x)\, dx$ is what's called a definite integral, and it's got lots of applications. For example, if a function is positive,

$\int_a^b f(x)\,dx$ gives exactly the area under the curve $y = f(x)$ for x between a and b. The amazing fact, which is so amazing it is called the Fundamental Theorem of Calculus, is that $\int_a^b f(x)dx$ can be found, like I said, by taking the indefinite integral and evaluating it at the top limit of integration, and then subtracting the value of it at the bottom limit. For example,

$$\int_1^2 x^2\,dx = \frac{1}{3}x^3 \bigg]_1^2 = \frac{1}{3}2^3 - \frac{1}{3}1^3 = \frac{7}{3}.$$

"You can even have

Improper Integrals.

That's when some of the limits are infinite. Then

$$\int_a^\infty f(x)\,dx = \lim_{b\to\infty} \int_a^b f(x)\,dx.\text{"}$$

Connor dropped some bullets in a baggie.

"Okay," he sighed. "That's a lot of stuff. But I think I get enough to understand Dad now."

"No you don't," said Ellie. "Lots of times, Dad uses

Multivariable Calculus."

"Multiwhatable?" said Connor.

"Multivariable. So far we have been talking about single variable calculus. You know, we have only had one variable, usually t or x. All of our functions have depended on only one variable."

"What's wrong with that?"

"It's unrealistic. In the real world, there are lots of variables. Think about predicting the weather. You need to know temperature, humidity, wind patterns, other nearby weather."

"Even then, you can't really predict the weather."

"Yes, but my point is that most things depend on a whole lot of other things, so there are multiple variables. So then you need to know about multivariable calculus."

"How hard can that be? You just do the same thing twice, or three times..."

"It's not that simple. For instance, in integration, you can have

Double Integrals."

"I just said that. You just do integration twice."

"Yes, but setting them up is not so easy. Suppose we want to integrate a function $f(x, y)$ over a region R in the xy-plane. We then take a double integral $\int \int f(x, y) dA$, where dA can be $dxdy$ or $dydx$. We choose the limits of integration to make sure we're integrating over the right region R. The limits on the outer integral are always constants, but the limits on the inner integral can depend on the variable of the outer integral.

"So, for example, $\int_{-1}^{1} \int_{x^2}^{1} x^3 dy dx$ is the integral of x^3 over this region R" (Figure B.9).

"We do the inner integral, which is a y-integral, first."

$$\int_{-1}^{1} \int_{x^2}^{1} x^4 dy dx = \int_{-1}^{1} x^4 y \Big]_{x^2}^{1} dx$$

$$= \int_{-1}^{1} x^4 - x^6 \, dx = \frac{x^5}{5} - \frac{x^7}{7} \Big]_{-1}^{1}$$

$$= \frac{1}{5} - \frac{1}{7} - \left(\frac{-1}{5} - \frac{-1}{7} \right) = \frac{4}{35}$$

"Okay, so that's double integrals."

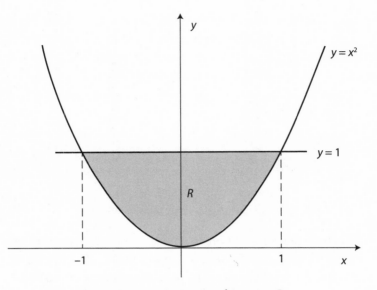

Figure B.9: Integrating over this region R.

"No, that's not all. You can also have double integrals in polar coordinates."

"Polar as in the North Pole?"

"No, polar as in a different set of coordinates. Usually, we use rectangular coordinates (x, y) to describe points in the plane. But we can also use polar coordinates (r, θ), like here" (Figure B.10).

"The coordinate r is just the distance from our point back to the origin. By the Pythagorean Theorem, which even you should know, $r = \sqrt{x^2 + y^2}$. The coordinate θ is just the angle between the line segment from the origin to the point and the positive x-axis.

"Sometimes we want to be able to convert a double integral in rectangular coordinates into one in polar coordinates. We have to choose the limits of integration so that

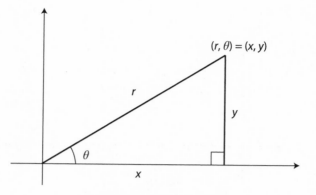

Figure B.10: Polar coordinates.

we cover the same region R. The interesting part is that the differential of area dA that began as $dy\,dx$ or $dx\,dy$ becomes $r\,dr\,d\theta$. The extra r comes from the fact that if we vary each of the two variables r and θ by a little bit, we get a region like this." She drew Figure B.11. "Its area is approximately $r\,d\theta\,dr$." (See p. 164 for more on this.)

"I'm tired of integrals,"said Connor as he dropped some more bullets in a baggie. "What are those things Dad always mentions? He calls them   **differential equations**. He acts like they are really important."

"They are really important. They describe everything."

"No, they don't. They don't describe what I'm thinking about you right now."

"I mean that whatever process you're interested in, whether it's the spread of zombies, or the growth of plants, or gravity, it can probably be explained using differential equations."

"So what's a differential equation?"

"It's just an equation that involves derivatives. For example, $\frac{dx}{dt} + 3x^2 - 7 = 0$ is a differential equation. So is $\frac{d^2y}{dx^2} + 3\frac{dy}{dx} - 6y + x^2 = 4$."

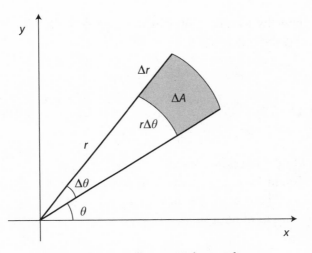

Figure B.11: A small area in polar coordinates.

"Okay, so what is something described by a differential equation?"

"Well, take $\frac{dy}{dt} = ky$. This is a very common differential equation, because it just says that the rate of growth of a population is proportional to the size of the population. The bigger the group, the faster it grows. Think of bacteria in a very big petri dish with lots of nutrients. The bacteria are happy, they have all their needs met, so they procreate."

"By procreate, do you mean have sex?"

"No, I don't. Bacteria don't have sex. They just split to make more bacteria. And the more of them there are, the more new ones that are produced."

"But what good is the differential equation?"

"You have to solve it for it to be useful. To solve this equation, you use **separation of variables**. In this case, we can do some algebra to separate all the terms that involve y on the left and all the terms that involve t on the right. So we divide by y and multiply by dt to get $\frac{dy}{y} = kdt$. Then we integrate both sides.

$$\int \frac{dy}{y} = \int kdt$$

This gives us

$$\ln|y| = kt + C."$$

"I told you I don't like ln."

"Tough luck. Deal with it. We're going to get rid of it anyway. Since e^x is the inverse function of $\ln x$, we take exponentials of both sides, and get

$$y = e^{kt+C},$$

$$y = e^{kt}e^C = y_0 e^{kt},$$

by letting $e^C = y_0$. So our solution is $y = y_0 e^{kt}$." (See p. 19.)

"Okay, I think I've seen enough calculus for one day," said Connor as he dropped the bullets in his hand back into the bowl, and stood up.

"What do you mean? We're almost done."

"We are?"

"Yeah. I've told you almost everything you need to know."

"You have?"

"Yup, just need to explain vectors. Sit down and keep sorting."

Connor sat back down.

"So, now," said Ellie. "Let's talk about

Vectors."

"What's a vector?"

"A vector is just an arrow, with a particular direction and a particular length."

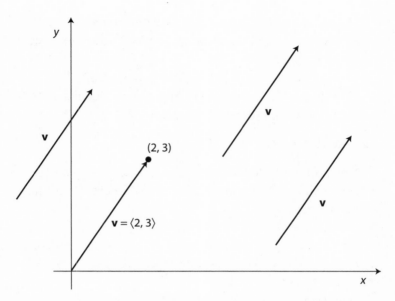

Figure B.12: The vector $\mathbf{v} = \langle 3, 2 \rangle$. It has direction and length.

"We shot arrows in gym once. Aaron Menand almost shot Kyla Devereaux. Missed her by an inch. He got in a lot of trouble."

"From what I heard about him, he used to get in a lot of trouble all the time."

"Yeah, the gym teacher should have known better than to give him a bow and arrow."

"Anyway, back to vectors. The initial point of the vector can be anywhere. We describe a vector in the plane by giving its x and y coordinates. So we might write $\mathbf{v} = \langle 3, 2 \rangle$. If we draw the vector so that it starts at $(0, 0)$, then this vector will end at the point $(3, 2)$. But remember, we can start it anywhere." (See Figure B.12.)

"By the Pythagorean Theorem, the length of $\mathbf{v}$, denoted $|\mathbf{v}|$, is given by $|\mathbf{v}| = \sqrt{3^2 + 2^2} = \sqrt{13}$."

"Okay, so we have these arrows, but what good are they?"

"They're good for lots of things. In physics, they get used all over the place."

"I don't care about physics. We're doing calculus."

"Physics is calculus, or a lot of it is anyway. But okay, first let me tell you about the next thing:

Parametric Curves.

"Suppose we're looking down on the science quad and we're watching the provost, Mildred Wilshore, sprinting across the quad."

"What's a provost?"

"It's someone at the college who worries a lot about money. You know Claire Wilshore in eighth grade? She was the provost's kid."

"Oh yeah."

"So imagine first, the provost veers left to avoid one zombie and then she veers right to avoid another. So her path looks like this" (Figure B.13).

"How can we keep track of that path?" continued Ellie. "The easiest way is to keep track of the point (x, y) on that path as time progresses. But since both her x-coordinate and her y-coordinate are changing, we give each coordinate as a function of time t. So we write $x = f(t)$ and $y = g(t)$. Then as time progresses, we get her coordinates at any given time as $\langle f(t), g(t) \rangle$. We call this her **position function** and denote it

$$\mathbf{p}(t) = \langle f(t), g(t) \rangle.$$

"In the case of Provost Wilshore, we could use $\mathbf{p}(t) = \langle 10 - 4 \sin t, t \rangle$. The sine function makes the x-coordinate wave back and forth. This weird 'function,' which is really two functions stapled together, is called a *vector-valued*

218

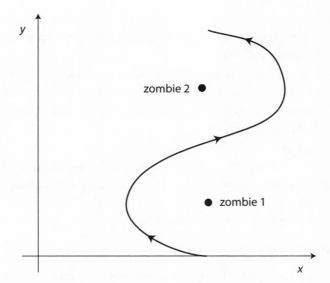

Figure B.13: The path taken by the provost to avoid zombies.

function , since for any value t that we put in, we get out a vector $\langle f(t), g(t) \rangle$."

"Why do you use the funny sharp vector brackets for it? Why not just use $(f(t), g(t))$?"

"It doesn't really matter. You could do that, too. But when we take derivatives of it, we'll want it to be a vector. So it makes sense to think of this as a vector now. But just think of the vector as starting at the origin and ending at $(f(t), g(t))$."

"Okay."

"Okay. So let's look at some simpler examples. Suppose instead, Provost Wilshore cuts across the quad with position function $\mathbf{p}(t) = \langle t, t \rangle$. Then as t varies from time $t = 0$ to time $t = 30$ seconds, since $x = t$ and $y = t$, it must be that $x = y$ the whole time, so she takes the straight line diagonal."

219

"Sure."

"Okay, now suppose instead, she takes the path $\mathbf{p}(t) = \langle t, t^2 \rangle$. What will that look like?"

Connor took the pen from Ellie. "Well, since $x = t$ and $y = t^2$, then for all t, $y = x^2$. We talked about curves like that in Mr. Thistle's class. That's a parabola." He drew the parabola on the paper.

"Good!"

"So I get this. Parametric curves can be used to describe whatever path someone is taking. But what does that have to do with calculus?"

"Calculus comes in when we take

Derivatives of Parametric Curves."

"How do you take the derivative of a function that has two parts $\langle f(t), g(t) \rangle$?" asked Connor.

What do you think?

"You just take the derivative of each part."

"Exactly. We take the derivative of $\mathbf{p}(t)$, which we call the velocity vector $\mathbf{v}(t) = \langle f'(t), g'(t) \rangle$. This is a new vector-valued function. The big question is, 'What does it tell us?' "

"I don't know. What does it tell us?"

"It points in the direction of motion. If you place the vector $\mathbf{v}(t)$ so it starts at the point on the curve where the provost is at that moment, it points in the direction she is headed at that moment." (See p. 166 for more on this.)

"Okay."

"And its length is the speed at that instant."

"What do you mean?"

"The length of the velocity vector, which is given by $|\mathbf{v(t)}| = \sqrt{(\mathbf{f'(t)})^2 + (\mathbf{g'(t)})^2}$, is in fact the speed with which the provost is running at that instant. So for example, if Provost Wilshre has position function $\mathbf{p}(t) = \langle t, t \rangle$, then

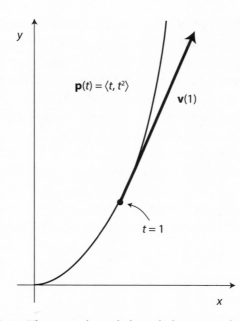

Figure B.14: The provost's parabolic path showing a velocity vector.

since $df/dt = 1$ and $dg/dt = 1$, $\mathbf{v}(t) = \langle 1, 1 \rangle$, so the velocity vector points in the direction of the diagonal line along which she is running and its length is $|\mathbf{v}(t)| = \sqrt{1^2 + 1^2} = \sqrt{2}$. So she is running at a constant speed of $\sqrt{2}$ feet per second. Probably not fast enough. . . "

"But if she runs with position vector $\mathbf{p}(t) = \langle t, t^2 \rangle$, then $\mathbf{v}(t) = \langle 1, 2t \rangle$. Notice that as time passes, this vector changes both direction and length. But at all times, it is tangent to this parabolic path, like in this picture."(See Figure B.14.)

"Its length is given by $|\mathbf{v}(t)| = \sqrt{1^2 + (2t)^2} = \sqrt{1 + 4t^2}$. So initially, at time $t = 0$, she is only going 1 foot per second. But notice that as time progresses, she is speeding up. By time $t = 1$ second, she is going $\sqrt{5}$ feet per second, and by $t = 2$ seconds, she is going $\sqrt{17}$ feet per second.

221

Obviously, this is not a very realistic model, since by time $t = 20$ seconds, she will be going $\sqrt{1601} \approx 40$ feet per second, faster than any human has ever travelled on foot."

"That's kind of cool that you can figure out her speed and direction from the parametric curve equations."

"Yeah, it is. So that's it. Enough calculus so you can understand when Dad starts talking about it."

"Thanks." Connor stood up.

"Wait a minute. We're not done sorting these bullets. Sit back down."

"It's not my job. It's yours," said Connor. He grinned as he waved and walked away, leaving Ellie with the half-filled bowl of bullets.

Acknowledgments

I owe a great deal of thanks to a lot of people, without whose help this book would not have come to be. Thanks to Alexa and Colton Adams, who gave me lots of ideas, intentionally and otherwise. I would go through the apocalypse with them anytime.

Thanks to Vickie Kearn, my editor at Princeton University Press. Many editors from other publishers said, "Oh, no, we wouldn't ever do a book like that, but you should talk to Vickie Kearn." Vickie was enthusiastic right from the start, and she has been incredibly supportive throughout the process. Quinn Fusting, Glenda Krupa, Ali Parrington, and all the rest of the team at Princeton University Press have been great to work with.

In writing this book, I took advantage of various books and articles, a list of which appear in the bibliography. I especially wanted to single out Paul Nahin's *Chases and Escapes* which was particularly helpful with regard to the path a zombie takes when pursuing a human.

I asked many friends, colleagues, and students to read the book and give me feedback. I am very grateful to them both for their support of this project and for their very helpful suggestions. These include Richard Bedient, Edward Burger, Michael Cestone, Eugene Choe, Satyan Devadoss, Tayana Fincher, Thomas Garrity, Meredith Greer, Joel Hass,

Ashwin Lall, Susan Loepp, Lew Ludwig, Steve Miller, Frank Morgan, Amy Myers, Daniel Packer, Noah Sandstrom, Theodore Sandstrom, Laura Taalman, Ron Taylor, and Robert Thistle. I am especially grateful to Courtney Gibbons for her careful reading and numerous useful suggestions.

As always, I want to thank Williams College and the Mathematics and Statistics Department at Williams College. The college and the department are incredibly supportive and both provide a wonderful environment that fosters creativity. I am also grateful to all my students, from whom I learn every day.

Unfortunately, I have no idea who among those I have mentioned above have managed to survive the zombie onslaught. But given their collective expertise, I have every hope that quite a few of them still roam this earth.

Bibliography

[1] Adams, Colin, Hass, Joel, Thompson, Abigail, How to Ace Calculus: The Streetwise Guide, W. H. Freeman, (1998)

[2] Adams, Colin, Hass, Joel, Thompson, Abigail, How to Ace the Rest of Calculus: The Streetwise Guide, W. H. Freeman, (2001)

[3] Anderson, Roy, May, Robert, Infectious Diseases of Humans: Dynamics and Control, Oxford University Press, (1992)

[4] Edelstein-Keshet, Leah, Mathematical Models in Biology, Society for Industrial and Applied Mathematics, (2005)

[5] Munz, Philip, Hudea, Ioan, Imad, Joe, Smith?, Robert J., When Zombies Attack!: Mathematical Modelling of an Outbreak of Zombie Infection, In: Infectious Disease Modelling Research Progress, ed. by Tchuenche, J. M. and Chiyaka, C., Nova Science Publishers, Inc., (2009) 133-154

[6] Murray, James D., Mathematical Biology, 3rd edition, Springer Verlag, (2007)

[7] Nahin, Paul, Chases and Escapes, Princeton University Press, (2007)

[8] Nowak, Martin, Bangham, Charles, Population dynamics of immune responses to persistent viruses, Science, new series 272, Issue 5258 (1996) 74-79

[9] Rogawski, Jonathan, Adams, Colin, Calculus, 3rd edition, W. H. Freeman, (2014)

[10] Wodarz, Dominik, Nowak, Martin, Mathematical models of HIV pathogenesis and treatment, Bioessays 24 (2002) 1178-1187

Index

68-95-99.7 Rule, 43

acceleration, 34, 205; average, 34; gravity, 51

basic reproductive rate, 104
blood brain barrier, 97, 99
brain cells: equilibrium value, 101

cell: average life span, 102, 179; endothelial, 97, 99
circle pursuit, 173
concavity, 205

definite integrals, 210
derivative, 16; definition of, 166, 201; second, 34, 205
differential, 207
differential equation, 23
differentiation: $\ln x$, 204; e^x, 204; Chain Rule, 203; Constant Multiple Rule, 202; Constant Rule, 201; Power Rule, 202; Product Rule, 203; Quotient Rule, 203
double integrals, 212; in polar coordinates, 164

encephalitis, 97
exponential decay, 115, 141
exponential growth, 19

force, 34, 159; maximum, 159

hippocampus, 99

impulse, 160
inflection point, 205
integrals, 206
integration: e^x, 208; Constant Multiple Rule, 208; Constant Rule, 208; Power Rule, 208; Sum Rule, 208
integration by parts, 180, 209

L'Hôpital's Rule, 181
limits of integration, 210
logistic growth, 81, 175
Lotka-Volterra model, 139

multivariable calculus, 211

Newton's Law of Cooling, 113
normal curve, 40
normal distribution, 42, 162; standard, 41

parametric curves, 218, 220
polar coordinates, 163
position function, 57, 218
predator-prey model, 139, 184
pursuit, 55, 169; circular, 68

radiodrome, 59
recurrence relation, 155

separation of variables, 215
slope field, 185
standard deviation, 42
statistics, 40

tangent line, 56, 194
tangent vector, 56, 166

u-substitution, 100, 208

vectors, 216
vector-valued function, 219
velocity, 34, 205
virion, 96
virus: capsid, 96; cytopathic, 99;
    HIV, 99; rabies, 97; West Nile,
    99